Aids to Anatomy

Aids to Anatomy

by Jack Joseph MD, DSc, FRCOG

Professor Emeritus, University of London,
formerly Head of Department of Anatomy,
Guy's Hospital Medical School, London and
Examiner in Anatomy for MB, BS and BDS,
London University and for the Primary Fellowship
of the Royal College of Surgeons of England;
presently Examiner for the Primary Fellowship
of the Royal College of Surgeons of Edinburgh
and the Royal College of Physicians and
Surgeons of Glasgow

Thirteenth Edition

Baillière Tindall · London

Published by Baillière Tindall
1 St Anne's Road, Eastbourne
East Sussex BN21 3UN

© 1984 Baillière Tindall

First edition 1876
Second edition 1882
Third edition 1889
Fourth edition 1893
Fifth edition 1902
Sixth edition 1907
Seventh edition 1913
Eighth edition 1920
Ninth edition 1933
Tenth edition 1940
Eleventh edition 1951
Twelfth edition 1962
Thirteenth edition 1984

During part of its long history, this book appeared as 'The
Pocket Gray' and as 'The Pocket Anatomy', by which titles it is still
widely known.

The tenth, eleventh and twelfth editions were revised by
Professor R.J. Last.

Typeset by Scribe Design Ltd, Gillingham, Kent
Printed and bound in Great Britain at University Press,
Cambridge

British Library Cataloguing in Publication Data

Joseph, Jack, 19__–
  Aids to anatomy—13th ed.
  1. Anatomy, Human
  I. Title   II. Last, R.J.
  611   QM23.2

  ISBN 0 7020 0960 1

Contents

Preface to the 13th Edition

A book which first appeared in 1876 and has had 12 editions and 26 reprints has proved its popularity with and value to many generations of medical students. In the 13th edition the basic and obviously successful format has been retained, and it remains a book on systematic anatomy without a section on osteology.

Much of the text has been rewritten and rearranged with parts added, especially on functional anatomy in relation to the joints. Many of the tables of relations of different structures have gone and have been replaced in a shortened form more suitable for modern teaching.

All the figures have been redrawn with standardized labelling. A few have been removed and 16 new figures added. Although the last edition introduced much of the new terminology, subsequent changes have made it necessary to bring this up to date, but where older names are still being used these have also been given.

Great care has been taken to make the headings and subheadings coherent and to use a clear typeface so that students can read the descriptions of the structures easily and solve the problem of remembering what they have tried to learn. Attitudes to topographical anatomy keep changing in medical education, but no one can deny the necessity and importance of a knowledge of the structures of the human body for understanding many aspects of medicine.

I would like to thank the publishers, and especially David Dickens and Cliff Morgan, for their cooperation in producing this new edition which I hope will continue to help the heavily burdened medical students over the early hurdles in their chosen profession.

Jack Joseph
London, August 1983

1
The Joints

A *joint* is the arrangement whereby separate bones or cartilages are attached to each other. Different kinds of joints can be classified either structurally or functionally.

Structural classification (Fig. 1)

Fibrous joints
The articulating surfaces are united by fibrous tissue continuous with the periosteum or perichondrium. Movement depends on the amount of fibrous tissue between the bones and varies from nil (e.g. sutures of adult skull) to a wide range (e.g. interosseous membrane of forearm).

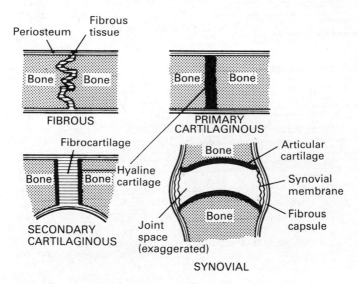

Fig. 1 Structural classification of joints (all joints are in section).

Cartilaginous joints

Primary cartilaginous: Bones are united by hyaline cartilage (e.g. the union of an epiphysis with the diaphysis). There is no movement.

Secondary cartilaginous (symphysis): Each articulating bone surface is covered with hyaline cartilage. The two cartilaginous plates are united by fibrocartilage which often contains a cavity. The cavity contains either a fluid like lymph (symphysis pubis, manubriosternal joint) or a gel (intervertebral disc). It never contains synovial fluid. There may be a limited amount of movement.

Synovial joints

Free surfaces covered with hyaline cartilage (except joints of clavicle and mandible). Enclosed in fibrous capsule. Capsule and all non-articulating surfaces lined by synovial membrane. Cavity contains a viscous synovial fluid secreted by the membrane. Movement varies from very little (e.g. sacro-iliac joint) to great freedom (e.g. shoulder, hip). Classified according to shape of bone surfaces e.g. *ball and socket, ellipsoid, condylar, saddle-shaped*, or type of movement, e.g. *hinge, pivot, gliding* (also called *plane*).

Functional classification

Freely movable joints—most synovial.
Slightly movable joints—symphyses. (These are permanent joints of movement.)
Immovable joints—*sutures* and *synchondroses*. (These are temporary joints of growth.)

In describing a joint the following systematic order is recommended:
(1) definition, type and articulating surfaces
(2) fibrous capsule and ligaments
(3) synovial membrane and bursae, if synovial joint
(4) nerve and blood supply
(5) movements and muscles producing the movements
(6) stability: factors are (a) bony, (b) ligamentous, (c) muscular
(7) important relations.

Development of joints

Condensations of mesoderm may ossify directly (*ossification in membrane*—e.g. flat bones of skull) or may chondrify first (*ossification in cartilage*—e.g. base of skull, long bones generally). Mesoderm between bones produces joint structures. Joint cavity results from disappearance of mesoderm. Perichondrium persists around joint cavity as capsule lined with synovial membrane. Cartilage persists over bone ends as articular cartilage.

During growth nutrient artery of shaft supplies down to cartilaginous growth plate. These are end-arteries. Epiphysis and capsule supplied by circulus vasculosus (of Hunter), lying between attachments of capsule and synovial membrane. No anastomosis between circulus vasculosus and nutrient artery of shaft until ossification of growth plate, at completion of growth.

Nerve supply of joints

Nerve to a muscle commonly gives sensory branch to joint on which the muscle acts, and to skin over joint (Hilton's Law). Nerves are associated with pain or movement (mechanoreceptors).

THE JOINTS OF THE VERTEBRAL COLUMN

Before birth whole column is concave forwards (*primary curve*). At about six weeks the head is held up and at about six months sitting up occurs. These produce forward convexities in the cervical and lumbar regions respectively. The latter becomes more marked on standing at about one year. These are called *secondary curves*. Thorax and sacrum retain primary curves.

The bodies of the vertebrae are united to one another by *intervertebral discs* which yield sufficiently to allow forward, backward and lateral bending and rotation of the column. This involves movement between the vertebral arches, so that there are synovial joints on each side between each pair of arches.

Joints of the bodies (symphyses)

These are secondary cartilaginous joints. The articulating surfaces are flat, except in the neck, where their edges are reciprocally bevelled and lipped and form small synovial joints (Fig. 2a).

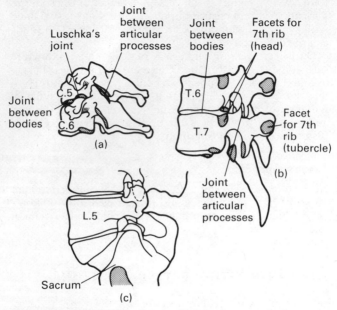

Fig. 2 Typical cervical, thoracic and lumbar vertebral articulations.

Intervertebral discs

The upper and lower surfaces of a vertebral body are covered by a plate of hyaline cartilage. The *anulus fibrosus*, uniting adjacent plates, is made of concentric layers of fibrous tissue. In each layer the fibres are parallel and slope at 45°. Alternate layers are at right angles to each other throughout the anulus. Slightly posterior to true centre of anulus is a gel, the *nucleus pulposus*, a remnant of the notochord. The gel resists loss of volume by pressure of body weight, and centrifugal thrust of nucleus tends to stretch fibres of anulus.

Backward displacement of the nucleus pulposus or of injured disc tissue, causing pressure on the spinal cord or nerves, is a recognized clinical entity.

The *ligaments* are *anterior* and *posterior*, the more superficial fibres forming a continuous band. The anterior longitudinal ligament extends from anterior arch of atlas to sacrum, and the posterior longitudinal ligament from the body of the axis to the sacrum. Anterior longitudinal ligament broadens from above downwards and is adherent to periosteum of each vertebral body. Posterior longitudinal ligament is attached more firmly to discs than to vertebrae, and is wider opposite discs than opposite vertebrae.

Joints of the arches (zygapophyseal joints)

These are gliding synovial joints between the inferior articular processes of the vertebra above and the superior articular processes of the vertebra below.
The facets are arranged as follows (Fig. 2):

Region	Shape and direction	Movement allowed
Cervical	Flat. In same plane.	Flexion and extension.
	Inferior process downwards and forwards.	Lateral flexion. Slight rotation.
	Superior process upwards and backwards.	
Thoracic	Flat. On arc of circle.	Limited flexion and extension.
	Inferior process forwards.	Rotation.
	Superior process backwards.	Slight lateral flexion.
Lumbar	Inferior process convex and faces laterally.	Flexion and extension and lateral flexion.
	Superior process concave and faces medially.	Limited rotation.

Movements. Head nodding (flexion and extension) and rotation occur at *atlanto-occipital* and *atlanto-axial joints.*

Elsewhere in column flexion and extension and lateral flexion are possible in all regions (range of each varies with region). Pure rotation possible only in thorax (limited by splinting effect of ribs). Lateral flexion (bending) in neck and thorax (not in lumbar region) accompanied by secondary rotation so that tip of spinous process rotates towards lateral convexity of curve (seen in scoliosis—lateral curvature of the spine).

Ligaments: (1) Capsular ligament of the above joints. (2) The laminae are connected by the *ligamenta flava*—thick, longitudinal, elastic bands attached above to the anterior aspect of the lower border of a lamina, and below to the posterior aspect of the upper border. (3) The spines are connected by weak *interspinous ligaments*, and their tips by a stronger *supraspinous ligaments*. In the neck, the *ligamentum nuchae* is a modification of these. It is not a stout elastic band in man, as it is in some quadrupeds, but a thin sickle-shaped intermuscular septum attached above to the occipital crest and below to the spinous process of the 7th cervical vertebra. Its deep edge reaches the spinous processes of the other cervical vertebrae; its superficial edge reaches the investing fascia over the posterior neck muscles. (4) *Intertransverse ligaments*, weak and thin, connect the transverse processes.

Special joints associated with movements of the head

The 1st cervical vertebra (atlas) articulates above with the occipital bone, and the joints between the atlas and axis (2nd cervical vertebra) are greatly modified to allow rotation of the head. These joints are in series with the lateral joints between the bodies of the cervical vertebrae (p. 4), not with the zygapophyseal joints of other vertebrae. The 1st and 2nd cervical nerves emerge behind the atlanto-occipital joints and lateral atlanto-axial joints respectively.

Atlanto-occipital joints (Fig. 3)
The articular surfaces are the upper, reniform, concave surface of the lateral mass of the atlas, and the convex occipital condyles. They are connected by a capsule. The

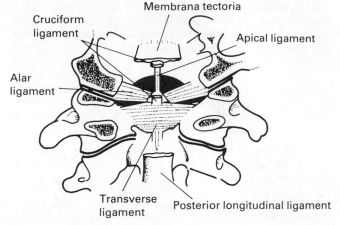

Cruciform ligament

Membrana tectoria

Apical ligament

Alar ligament

Transverse ligament

Posterior longitudinal ligament

Fig. 3 Ligaments connecting the skull to atlas and axis.

anterior and posterior arches of the atlas are connected to the margins of the foramen magnum by the *anterior* and *posterior atlanto-occipital membranes*. The lower edge of the posterior atlanto-occipital membrane arches across the groove on the posterior arch medial to the lateral mass of the atlas. The vertebral artery with the 1st cervical nerve between it and the bone lies in this groove.

Movements: Flexion and extension. Lateral flexion. No rotation.

The stability of the atlanto-occipital joints depends on the skull being held down by ligaments connecting the skull to the axis, and by short muscles surrounding the joints (p. 67 and Fig. 26).

Atlanto-axial joints (Fig. 3)

(1) *Median:* The *transverse ligament* of the atlas, attached to tubercles on the medial aspect of the lateral masses, shuts off a small anterior compartment for the dens from a larger posterior containing the spinal cord.

The ligament lies in a groove on the dens. The dens projects into the anterior compartment and has a smooth cartilaginous surface in front to form a synovial joint with a facet on the back of the anterior arch of the atlas. Similarly the dens is covered by cartilage behind and articulates with the front of the transverse ligament by an intervening bursa.

(2) *Lateral:* At the junction of pedicle with body the axis has a large oval facet facing upwards and slightly outwards. The inferior surface of the lateral mass of the atlas has a similar facet articulating with this by a synovial joint.

Movements: When the atlas rotates, carrying with it the head, the lateral mass slides forwards and backwards at this joint.

Running downwards and inwards from the back of the lateral mass to the back of the body of the axis is a thin but strong *accessory atlanto-axial ligament.*

Other ligaments connecting the skull to the axis: (1) The *apical ligament* from the tip of the dens to the anterior margin of the foramen magnum (remnant of the notochord). (2) The strong *alar* (*check*) *ligaments* connecting the sides of the dens to the medial side of the occipital condyles. (3) Bands stretching up from the back of the body of the axis to the transverse ligament and thence to the foramen magnum. These with the transverse ligament form the *cruciform ligament*, of which they are the superior and inferior bands. (4) The *membrana tectoria* is a broad band separating the cruciform ligament from the dura mater of the spinal cord—attached to the body of the axis with the posterior longitudinal ligament. Above it enters the skull through the foramen magnum to fuse with the periosteum of the basi-occiput. The spinal dura mater is adherent to the membrana tectoria.

Lumbosacral joints (Fig. 2)

The inferior surface of the 5th lumbar vertebra faces downwards and backwards, its body being a wedge narrow behind. This wedge supports the whole weight of the trunk above, and has a tendency to slip forwards (*spondylolisthesis*). This is prevented by the more transverse direction of the joints between the inferior articular processes of the 5th lumbar vertebra and the superior articular processes of the sacrum, the thick intervertebral disc and the strong *iliolumbar ligaments*, from transverse process of 5th lumbar vertebra, backwards and downwards as well as laterally to iliac crest (Fig. 10).

Note: Vertebral joints are supplied by branches of the corresponding spinal nerves. These lie in the intervertebral foramen close to the lateral part of the joints between the bodies, the zygapophyseal joints and, in the cervical vertebrae, the small joints between the lateral edges of the bodies. Disease of any of these joints may affect the spinal nerve.

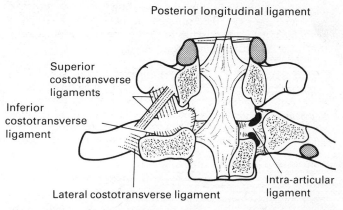

Fig. 4 Ligaments connecting rib to transverse process of a vertebra, joint cavity of head of rib and posterior longitudinal ligament.

JOINTS OF THE RIBS (Figs. 2 and 4)

A typical rib articulates posteriorly by its head and tubercle with the vertebral column; anteriorly the first seven costal cartilages articulate with the sternum, and the 8th, 9th and 10th with the costal cartilage above.

Joints of the heads of the ribs

From the 2nd to the 9th rib the head articulates with facets on the upper border of the corresponding vertebral body and the lower border of the vertebra above by a synovial joint. Between the two a ridge on the head of the rib is attached to the edge of the intervertebral disc by an *intra-articular ligament*. The anterior part of the capsule of the joint is reinforced by a *radiate ligament*, whose central limb crosses the midline. The first and the lowest three ribs articulate with only their own vertebral body.

Joints of the tubercles of the ribs (costotransverse)

The medial facet of the tubercle of ribs 1–10 articulates with a facet on the front of the corresponding transverse process by a synovial joint. The tubercle of the 11th and 12th ribs are attached by fibrous tissue to the transverse process of their corresponding vertebrae.

Costotransverse ligaments: (1) *Inferior costotransverse ligament* from back of neck of rib to front of transverse process (also called ligament of neck). (2) *Lateral costotransverse ligament* from lateral facet of tubercle to tip of transverse process. (3) *Superior costotransverse ligament* from upper edge of neck to lower border of transverse process above.

Joints of the costal cartilages with the sternum

First costal cartilage fixed by primary cartilaginous joint to manubrium; the two move as one and produce movement at the *manubriosternal joint*, a secondary cartilaginous joint which however synostoses as a rule after 40 years of age. Rigidity of 1st costal cartilage required for attachment of costoclavicular ligament to stabilize clavicle (p. 15) and thus whole of upper limb.

Segments of sternum (*sternebrae*) ossify separately. Costal cartilage articulates with upper and lower edges of adjacent sternebrae by synovial joint, divided into upper and lower compartments by small intra-articular ligament. As sternum ossifies, fusion of adjacent sternebrae is accompanied by disappearance of intra-articular ligament. Hence in adult chondrosternal synovial joints are single cavities, except for the 2nd, where intra-articular ligament persists.

Summary

First chondrosternal articulation non-synovial throughout life. Second a bilocular synovial cavity throughout life. Third to 7th synovial cavities bilocular in infancy, unilocular after ossification of body is complete in adolescence, and tend to become obliterated in old age. Tips of 8th, 9th and 10th costal cartilages form synovial joints, each with costal cartilage above it.

Movements of the thorax in respiration

The ribs are pulled up possibly by the intercostal muscles and, because of their curves, increase the transverse diameter of the thorax; also, since they are directed obliquely downwards their elevation increases the anteroposterior diameter of the chest and pushes the sternum forwards. This is the *pump-handle movement*. This movement takes place about an axis through the head and tubercle of the upper ribs whose costotransverse joints are curves. The lower ribs move outwards in a *caliper movement* about a vertical axis through the neck of the rib. Their costotransverse joints are flat and the result is an increase in only the transverse diameter of the thorax.

TEMPOROMANDIBULAR JOINT (Fig. 5)

A synovial, condyloid joint between head of mandible, and mandibular fossa and articular tubercle on the inferior surface of temporal bone. Right and left joints are regarded as one joint. Each has a double joint cavity designed to allow the head to glide on the base of the skull and also to rotate round its long axis.

Articulating surfaces: Head of mandible with its long axis directed backwards and medially, is covered by fibrocartilage which extends on to back of condyle.

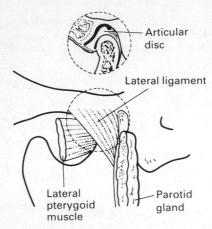

Fig. 5 Left temporomandibular joint seen from the side.

Mandibular fossa part of squamous temporal. The concave anterior surface of the tympanic plate is not part of the articular surface, and is occupied by parotid gland (see Fig. 5). The articular tubercle is covered with fibrocartilage, and is within the joint.

Capsule: Loose and superiorly attached anterior to the squamotympanic and petrosquamous fissures and in front of the articular tubercle. Below, the capsule is attached along the articular margin of the head of the mandible, and is strengthened laterally by the *lateral (temporomandibular) ligament*, whose fibres are directed from the posterior end of the zygomatic arch downwards and backwards to the neck of the mandible.

The *articular disc*, adapted to the bony surfaces, is attached all round to the capsule; it receives anteriorly fibres of the lateral pterygoid muscle which is also attached to the capsule.

Accessory ligaments: Two bands are usually described as additional ligaments of this joint: (1) *sphenomandibular ligament*, a flat band from spine of sphenoid to lingula and lower margin of mandibular foramen; (2) *stylomandibular ligament*, a fascial band from styloid process to posterior border of ramus.

Note: The sphenomandibular ligament is a part of the embryonic mandibular (or 1st pharyngeal) arch. This arch is completed by anterior ligament of malleus, malleus and incus above, and by Meckle's cartilage below. Note also that the maxillary vessels, auriculotemporal nerve and fibres of lateral pterygoid muscle separate the ligament from the mandible, but the chorda tympani nerve lies deep to the ligament.

Nerve supply: A relatively large branch from the auriculotemporal (mandibular division of trigeminal nerve) with a small contribution from the nerve to the masseter also from mandibular nerve.

Movements: The mandible (head and disc together) is protruded by the lateral pterygoid muscles and retruded by elastic recoil and the posterior fibres of temporalis muscles. The mouth is opened by the lateral pterygoid muscles which pull the head and disc forwards on to the articular tubercle, and by the digastric, geniohyoid and mylohyoid muscles which depress the chin, aided by gravity.

The mandible is pulled up by the combined action of the temporalis, masseter and medial pterygoid muscles (Fig. 22).

Side-to-side movements are produced by unilateral action of the pterygoids. Movement to the left involves rotation of the left condyle and pulling forwards of the right condyle by the right lateral pterygoid and possibly the right medial pterygoid muscles.

Note: The mandible moves round a transverse axis between the two lingulae. (This can be confirmed by finding an immobile point in one's own jaw, about half-way down the middle of the ramus.) If this point is immobile, the inferior dental nerve is not stretched during opening of the jaw.

The advantage of the gliding movements being added to the rotatory movements is that it allows grinding movements during mastication. It has the disadvantage that the condyle may slip beyond the articular tubercle and become fixed below the zygomatic arch (dislocation). This is not uncommon and occurs on only one side.

THE JOINTS OF THE UPPER LIMB

Sternoclavicular joint (Fig. 6)

A synovial gliding joint allowing a small amount of movement; this movement, however, takes place about a fulcrum near the joint so that there is considerable movement of lateral end of clavicle.

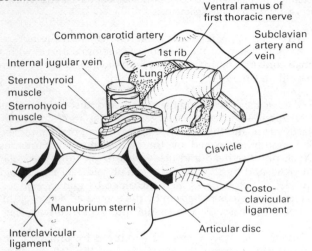

Fig. 6 Posterior relations of sternoclavicular joint.

Articulating surfaces: Medial end of clavicle with manubrium sterni and upper surface of 1st costal cartilage.

The superolateral angle of the manubrium sterni has a shallow cartilage-covered notch. The medial end of the clavicle, covered by fibrocartilage, is enlarged and prominent. It rests partly on this notch and partly on the 1st costal cartilage. There is no adaptation of the articular surfaces to one another.

In addition to the capsule, there are *anterior* and *posterior ligaments* and an *interclavicular ligament* which is very strong and stretches across the upper border of the manubrium. Outside the capsule the inferior part of the medial end of the clavicle is firmly bound to the first costal cartilage by the thick *costoclavicular (rhomboid)* ligament. This acts as the fulcrum of movement.

The joint is divided by an articular disc which, in addition to being attached by its periphery to the capsule, is attached below to the costal cartilage and above to the clavicle. Synovial membrane lines the capsule of each compartment of the joint.

Nerve supply: Supraclavicular nerves (C. 3, 4).

Movements: Clavicular movement is associated with scapular movement. In vertical and horizontal planes small amount of movement of medial end of clavicle around costoclavicular ligament becomes considerable movement in opposite direction of lateral end, upwards, downwards, backwards and forwards. Clavicle can also rotate (up to 40°) around its longitudinal axis.

Stability: Medial end of clavicle prevented from over-elevation by downward pull of disc and interclavicular ligament. Medial thrust from upper limb resisted by costoclavicular ligament, which is stronger than clavicle itself (clavicle fractures before ligament ruptures).

Posterior relations: Immediately behind the joint are the great vessels of the neck—on left, the left brachiocephalic vein and left common carotid artery; on right, the bifurcation of the brachiocephalic artery; between the vessels and the joint are the sternothyroid and sternohyoid muscles. Behind the vessels is the anterior border of the pleura.

Acromioclavicular joint

This is a synovial gliding joint allowing limited movement.

Articulating surfaces: Flat facet on clavicle faces downwards and laterally. Flat facet on acromion faces upwards and medially. The clavicle therefore overlaps the acromion, and if dislocated moves upwards and outwards.

The *coracoclavicular ligament* consists of two parts, *conoid* and *trapezoid*. The posterior conoid ligament, apex downwards, connects conoid tubercle of coracoid process to that of clavicle; helps muscles to resist downward pull of scapula and attached upper limb. Trapezoid ligament directed nearly horizontally outwards from top of coracoid to trapezoid ridge on clavicle; takes medial thrust of scapula from humerus to clavicle.

The joint also has an imperfect articular disc.

Nerve supply: Supraclavicular nerves (C. 3, 4).

Movements: Very little movement when scapula moves upwards, downwards, backwards and forwards. When upper limb is raised above shoulder, scapula rotates on clavicle so as to make glenoid cavity face upwards. The fulcrum of movement is the coracoclavicular ligament.

As upper limb moves forwards on chest wall with body of scapula, angle between coracoid and clavicle is reduced due to gliding at joint surfaces.

Note: (1) This joint is subcutaneous and easily palpable. (2) Upward and outward dislocation of outer end of clavicle is readily reduced, but will not remain reduced owing to inclination of surfaces.

Shoulder joint (Fig. 7)

This is a synovial ball-and-socket joint, allowing very free movement between the humerus and the scapula.

Articulating surfaces: Head of humerus forms nearly half a sphere and is directed medially, upwards and backwards.

Glenoid cavity of scapula is slightly concave and pear-shaped. It is much smaller than the head of the humerus. With the arm hanging at the side the cavity faces forwards as well as laterally. It is deepened slightly by a rim of fibrocartilage—the *glenoidal labrum*. Developmentally the cavity includes part of the coracoid process.

A secondary socket is formed above the joint by the coraco-acromial ligament, which with the coracoid process

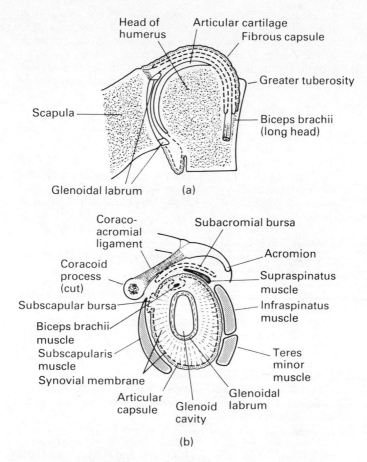

Fig. 7 Shoulder joint: (a) coronal section, (b) sagittal section.

and acromion forms the coraco-acromial arch. The coraco-acromial ligament is flat and triangular; its wide end is attached to the lateral border of the coracoid process, its apex to the acromion in front of the clavicular facet.

Capsule and ligaments: The capsule is very loose. It is attached to the scapula outside the glenoidal labrum, and to the humerus around the anatomical neck, except inferiorly where it extends about 2.5 cm down the shaft on its medial aspect. It bridges the upper end of the intertubercular sulcus, and it is perforated anteriorly by the subscapularis bursa.

The epiphyseal line of the humerus is extracapsular, except for a small medial part where the capsule extends down the shaft.

The synovial membrane lines the capsule and communicates with the subscapularis bursa. It encloses the tendon of the long head of biceps brachii and extends along the tendon deep to the transverse ligament (tendon is intracapsular). Medially it covers the small area of shaft within the joint.

(1) *Coracohumeral ligament*, from the coracoid process to the transverse ligament.

(2) *Glenohumeral ligaments*, three thickenings of capsule anteriorly—upper, middle and lower.

(3) Tendon of long head of biceps brachii passes from the supraglenoid tubercle across the top of the joint to the intertubercular groove.

(4) Tendons of infraspinatus, supraspinatus, subscapularis and teres minor are partly attached to capsule, which is strengthened above, in front and behind. Capsule with fused tendons is called *rotator cuff*. Rotator cuff is below coraco-acromial arch, separated by *subacromial bursa*.

Nerves from the suprascapular, axillary and subscapular.

Arteries from the suprascapular and anterior and posterior circumflex.

Movements: At glenohumeral joint—flexion, extension, abduction, adduction and lateral and medial rotation. Range of abduction and flexion greatly increased by movements of scapula on chest wall so that upper limb can be raised through 180° in both the coronal and sagittal planes. After about 40° of abduction or flexion, scapula rotates on chest wall simultaneously with movement of humerus on scapula. Supraspinatus required for initiation of abduction (to about 20°) to convert vertical pull of deltoid into a lateral rotatory movement. Rotation of scapula so that glenoid cavity faces upwards due to serratus anterior and lowest fibres of trapezius. Glenohumeral movement beyond 90° requires some lateral rotation. Of total 180°, movement at glenohumeral joint responsible for about 120°, rotation of scapula and clavicle together about 40° and rotation of scapula on clavicle about 20°.

Muscles producing flexion are anterior part of deltoid, clavicular head of pectoralis major, coracobrachialis. Adduction (limited to about 30°) due to pectoralis major, is possible only if limb flexed or extended from anatomical position. Extension (limited to about 40°) due to teres major, latissimus dorsi. Lateral rotation about long axis of humerus so that anterior surface of humerus faces laterally, about 50°; due to infraspinatus, teres minor. Medial rotation, anterior surface faces medially, about 50°; due to subscapularis, pectoralis major, teres major, latissimus dorsi.

Scapula can be pulled forwards (protraction) on chest wall increasing reach; due to serratus anterior. Also pulled backwards (bracing back the shoulders) due to trapezius, rhomboids. Elevated (shrugging shoulders) due to trapezius, levator scapulae. Clavicle accompanies scapula with movements at sternoclavicular joint (p. 15)

Stability: (1) Bony factors. Deepening of fossa by glenoidal labrum. Not important. (2) Ligamentous factors. Coracoacromial ligament prevents upward dislocation. (3) Muscular factors. Muscles of rotator cuff hold humeral head in glenoid cavity. Long head of triceps acts as a strap below the joint in full abduction. Dislocation is common. Usually downwards through lax part of capsule. Can injure axillary nerve which supplies deltoid, teres minor.

The elbow joint (Fig. 8)

This is a synovial hinge joint between the lower end of the humerus and the upper ends of the ulna and radius. Its cavity and some of its ligaments are common to it and the superior radio-ulnar joint, but the latter is described separately because of the different movement at this joint.

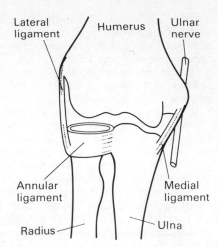

Fig. 8 Elbow joint opened from in front.

Articulating surfaces: The trochlear notch of the ulna articulates with the trochlea of the humerus (the medial edge of the humeral trochlea projects distally), the slightly concave circular upper surface of the head of the radius articulates with the convex capitulum of the humerus.

Capsule: Very thin in front and behind and is attached in front to the humerus above the coronoid and radial fossae and behind to the margins of the large olecranon fossa. Distally it is attached to the margins of the trochlea of the ulna and to the annular ligament of the superior radio-ulnar joint.

The epiphyseal line of the humerus is almost entirely intracapsular (the medial epicondyle is extracapsular), that of the ulna extracapsular, and that of the radius intracapsular.

Ulnar collateral ligament is triangular and attached proximally to the medial epicondyle and distally to medial margins of coronoid process and olecranon. Its distal edge is thickened, forming the transverse band, under which the posterior recurrent ulnar artery and a branch of the ulnar nerve enter the joint.

Radial collateral ligament runs from the lateral epicondyle to the annular ligament (it is not attached to the head of the radius, so as to allow this to rotate freely in pronation and supination) and the ulna.

Synovial membrane covers the floors of the olecranon, radial and coronoid fossae, and extends into the superior radio-ulnar joint, lining the annular ligament.

Nerve supply is from musculocutaneous, median, radial and ulnar.

Blood supply is from anastomosis around elbow-joint (profunda brachii, ulnar collateral, supratrochlear and radial, ulnar and interosseous recurrent arteries).

Movements: Flexion (brachialis, biceps brachii, brachioradialis) and extension (triceps brachii, anconeus).

Note: Carrying angle (lateral angle between humerus and bones of forearm) is due to medial part of lower end of humerus being lower than lateral. Matched by equal and opposite carrying angle at the wrist. Carrying angle of elbow obvious if forearm is extended and supinated; disappears in pronation (the working position) to bring axes of arm and forearm into line; also if forearm is flexed.

Important relations: The joint is covered by thick triceps brachii tendon behind, and by brachialis and brachioradialis in front. On brachialis are biceps brachii tendon, brachial artery and median nerve from lateral to medial. The ulnar nerve lying posteriorly on the medial epicondyle is in contact with the medial ligament.

Radio-ulnar joints

The head of the radius rotates in the radial notch of the ulna at the *proximal joint*; the lower end of the radius slides round the head of the ulna at the *distal joint*. The shafts of the bones and the interosseous membrane constitute the *middle joint*.

Proximal joint

The *annular ligament* (Fig. 8) is attached to the anterior and posterior margins of the radial notch of the ulna, and encircles the head of the radius. Its lower edge forms a smaller ring than its upper, so that the head of the radius cannot readily be dislocated downwards. The smaller radial head of a child can be more easily dislocated.

The *synovial membrane*, continuous with that of the elbow joint, is attached to the neck of the radius but bulges slightly below the annular ligament where it is covered by the *quadrate ligament*, which extends between the radius and ulna.

Distal joint
The convex head of the ulna is accommodated in the concave notch on the medial side of the lower end of the radius. The lower edge of the notch is attached to the lateral side of the base of styloid process of the ulna by a triangular disc of fibrocartilage, so that the styloid process more or less forms the centre of a circle round which the radius moves.

The joint has thin and lax anterior and posterior ligaments.

Synovial membrane, separate from that of wrist joint, bulges upwards deep to pronator quadratus (*recessus sacciformis*) anterior to interosseous membrane.

The fibres of the interosseous membrane are mostly directed downwards and medially, so that a force passing up the radius is transferred to the ulna. The *oblique cord* runs upwards from the lower part of the tubercle of the radius to the coronoid process; between it and the membrane the posterior interosseous vessels pass to the back of the forearm.

Pronation and supination
These are rolling movements of radius and hand about the ulna. In pronation the radius crosses in front of the ulna and the hand faces backwards (pronator teres and pronator quadratus). In supination the radius comes to lie parallel to the ulna and the hand faces forwards (supinator and biceps brachii). Axis of rotation may pass through centres of curvatures of head of radius and head of ulna if the ulna does not move. Almost always these movements are accompanied simultaneously by backward and lateral (in

pronation) or forward and medial movement (in supination) of lower end of ulna in order to centre the grip of the hand (e.g. turning a door knob, screwdriver, etc.). The axis may pass through the head of the radius and any of the digits.

The wrist (radiocarpal) joint (Fig. 9)

This is the synovial ellipsoid joint between the forearm bones and the proximal row of carpal bones. The slightly concave distal surface of the radius articulates with the

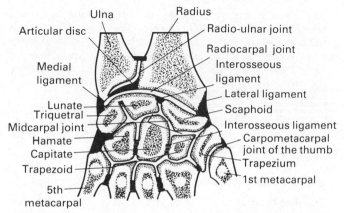

Fig. 9 Coronal section through distal radio-ulnar joint, and radiocarpal and intercarpal joints.

convex proximal surfaces of the scaphoid and lunate, and the triangular fibrocartilage separates the head of the ulna from the triquetral.

Ligaments: Anterior and posterior ligaments run medially and distally from the radius. Lateral and medial ligaments run from the styloid processes of radius and ulna respectively to carpal bones. All these also form ligaments of the carpal joints as well.

Movements: These are flexion and extension, and abduction and adduction. All these movements involve movements between the proximal and distal rows of the carpal bones. Flexion is produced by flexor carpi radialis, flexor carpi ulnaris and palmaris longus (if present). Extension is due to extensors carpi radialis longus and brevis and extensor carpi ulnaris. Abduction (radial deviation) is due to flexor carpi radialis and extensors carpi radialis longus and brevis, and adduction (ulnar deviation) to flexor carpi ulnaris and extensor carpi ulnaris. Adduction is much more extensive than abduction.

Carpal joints (Fig. 9)

Important features of these are interosseous ligaments which limit individual joint cavities. They occur (1) between the bones of the proximal row, shutting off the carpal joints from the wrist joint, and (2) between the distal row of bones, thus shutting off a large complex *midcarpal joint* from the carpometacarpal joints. These ligaments are not always complete, so that the joints may communicate.

The pisiform has an independent joint with the triquetral. Extensions of the tendon of the flexor carpi ulnaris form the *pisohamate* and *pisometacarpal ligaments*, the latter reaching base of 5th metacarpal.

Carpometacarpal joints

The first (that of the thumb) is an independent joint. The other four are subdivided by an interosseous ligament from the capitate to the 3rd metacarpal, so that there is a lateral joint between the trapezoid and capitate and the 2nd and 3rd metacarpals; and a medial joint between the hamate and the 4th and 5th metacarpals. These joints communicate with the joints between the bases of the four medial metacarpals. The joints between the metacarpals are limited distally by the *interosseous* (*deep transverse*) *metacarpal ligaments*. There is slight movement at the 4th and 5th carpometacarpal joints and none at the 2nd and 3rd.

Carpometacarpal joint of the thumb
This joint is independent of the others structurally and functionally. It is a synovial saddle-shaped joint between the trapezium and base of the 1st metacarpal, and has a

loose capsule. It allows flexion, extension, adduction, abduction and opposition. At rest, plane of thumb metacarpal is at right angles to remaining metacarpals so that flexion and extension are in the plane of the palm and abduction and adduction are at right angles to the palm. *Opposition* occurs when the tip of the thumb is brought into contact with the tip of any of the remaining digits and involves rotation of the thumb as well as flexion and adduction.

Flexion is due to flexor pollicis longus and brevis, extension to extensor pollicis longus and brevis, abduction to abductor pollicis longus and brevis and adduction due to adductor pollicis. Opponens pollicis, as well as the flexors and adductor, is used in the movement of opposition.

Metacarpophalangeal joints

These are synovial ellipsoid joints between the convex head of the metacarpal and the single oval concave facet on the proximal end of the proximal phalanx.

Anteriorly the capsule is thickened to form a fibrocartilaginous plate: posteriorly it is thin and lax. Thus flexion is free, but extension is usually limited to the straight position. The head of the metacarpal is wider in front than behind, and collateral ligaments, running forwards to the phalanx, are attached to the posterior part of the side of the metacarpal head. The result is that the digits can be moved from side to side in extension but not in flexion.

Interphalangeal joints

Each of these is a synovial hinge joint with well marked collateral ligaments. The head of the proximal and distal phalanges have two convex condyles and the base of the middle and distal phalanges have double concave facets.

Flexion and extension occur at the metacarpophalangeal and interphalangeal joints. Flexion is due to flexors digitorum profundus and superficialis and extension to the extensor digitorum (the interossei and lumbricals are also involved). The thumb has its own long and short flexors and extensors. Abduction and adduction at the metacarpophalangeal joints also take place relative to an axis through the middle digit. The dorsal interossei are the abductors and the palmar interossei are the adductors.

THE JOINTS OF THE PELVIS

The sacro-iliac joint (Figs. 10 and 11)

The sacro-iliac joint transmits the weight of the body through the pelvis to the lower limb. It is a synovial joint,

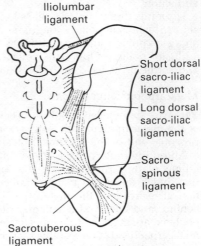

Iliolumbar ligament

Short dorsal sacro-iliac ligament

Long dorsal sacro-iliac ligament

Sacro-spinous ligament

Sacrotuberous ligament

Fig. 10 Ligaments of sacro-iliac joint seen from behind.

L.5

Fig. 11 Weight of body tends to tilt sacrum forwards; sacrospinous and sacrotuberous ligaments resist this.

but the articular surfaces (auricular surfaces of sacrum and of ilium), while coated with hyaline cartilage, are irregular and interlock with one another. Behind and above the articular surfaces are extensive rough areas connected by strong interosseous sacro-iliac ligaments, thus forming a fibrous joint. Note that the sacrum is wedge-shaped in coronal section at the level of the second sacral vertebra so that the sacrum is wedged into the pelvis.

Capsule and ligaments: In addition to the capsule there are (1) *interosseous sacro-iliac,* already referred to, (2) *anterior sacro-iliac,* weak and connecting anterior margins of articular surfaces, (3) *posterior sacro-iliac,* strong and connecting back of sacrum to ilium; lower part continuous with sacrotuberous ligament, (4) *sacrotuberous,* from back of sacrum to tuberosity of ischium, and (5) *sacrospinous,* anterior to sacrotuberous, from side of sacrum to spine of ischium.

Note: Ligaments (4) and (5) hold down the lower posterior part of the sacrum and resist the tendency of the weight of the body to rotate the upper part of the sacrum downwards and forwards.

Movements: Slight rotatory movements (5°) can occur. These may increase in pregnancy due to softening of ligaments.

Important relations: Passing down in front of the joint are the obturator nerve and the lumbosacral trunk and first sacral nerve. Also in front of the joint, separated by fat containing the nerves, are the bifurcation of the common iliac vessels, the veins being nearer to the joint. Ureter crosses pelvic brim in front of the bifurcation of the artery.

The symphysis pubis

The symphysis pubis is a secondary cartilaginous joint between the bodies of the pubic bones. The bony surfaces are covered by hyaline cartilage, and the cartilaginous surfaces are firmly united by an interpubic disc of fibrocartilage. The fibrocartilage may contain a small non-synovial cavity.

Ligaments are anterior, posterior, superior and inferior or subpubic (arcuate). The subpubic ligament is a thick band, continued above into the interpubic disc and having extensive attachment along the edges of the subpubic arch. It is separated from the perineal membrane (inferior layer of urogenital diaphragm) by the deep dorsal vein of the penis or clitoris.

No movements occur at this joint but some movement possible in pregnancy due to softening of ligaments.

THE JOINTS OF THE LOWER LIMB

The hip joint (Figs. 12 and 13)

This is a synovial ball and socket joint with a considerable range of movement. It differs, however, from the shoulder joint in that great stability is required to bear the superimposed weight. Accordingly, the head of the femur fits snugly into the deep acetabulum and the ligaments are strong. The freedom of movement of the thigh depends on the shaft of the femur being held away from the pelvis by the relatively long neck.

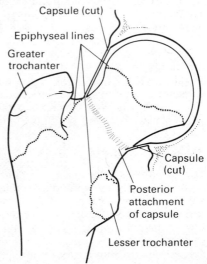

Capsule (cut)

Epiphyseal lines

Greater
trochanter

Capsule
(cut)

Posterior
attachment
of capsule

Lesser trochanter

Fig. 12 Diagram of
left hip joint opened
from behind.

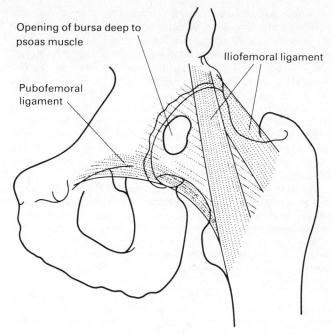

Opening of bursa deep to psoas muscle

Iliofemoral ligament

Pubofemoral ligament

Fig. 13 Left hip joint from in front.

Articulating surfaces: Acetabulum faces downwards and forwards as well as laterally. It is a cup-shaped fossa with prominent margins which are interrupted below at the acetabular notch. A large area extending from the notch to the centre of the acetabulum is not covered by articular cartilage, but is occupied by a pad of fat covered with synovial membrane. Rim of acetabulum is made more prominent by a fibrocartilaginous *acetabular labrum*; this forms a complete ring, bridging over the notch as the *transverse ligament*.

Head of femur forms two-thirds of a sphere; has small pit below and behind centre for attachment of *ligament of head of femur* (*ligamentum teres*) which is attached to the transverse ligament.

Capsule and ligaments: The capsule is strong and is attached to the hip bone outside the acetabular labrum and to the transverse ligament. Distally it covers the whole neck of the femur in front, extending to the intertrochanteric line; it leaves the lateral part of the neck uncovered behind, being attached to the neck 1 cm medial to the intertrochanteric crest. Small arteries passing up under the edge of the capsule to the neck and head of the femur carry with them fibrous bands, which form reflections of the capsule along the neck known as *retinacula.* Some of the deeper fibres encircle neck of femur and form *zona orbicularis.* The capsule may be perforated anteriorly where it communicates with the bursa deep to the psoas.

Iliofemoral ligament (Y-shaped ligament of Bigelow) spreads like an inverted Y downwards in front of joint from lower part of anterior inferior iliac spine to intertrochanteric line. Consists of very strong lateral and medial bands with thinner intermediate part inseparably blended with capsule. This ligament is very strong. It prevents hyperextension of the hip joint. (Note that the psoas bursa is medial to this ligament.)

Ischiofemoral ligament from ischium below and behind acetabulum, passes upwards and laterally over neck of femur and blends with capsule; many fibres encircle neck and form zona orbicularis.

Pubofemoral ligament on medial aspect of joint from iliopubic eminence to medial part of capsule.

Ligament of head of femur (ligamentum teres) is a flattened band which passes inside the acetabulum from the margins of the acetabular notch to the pit on the head of the femur. It is external to the synovial membrane which ensheaths it and it carries an artery to head of femur. Important as source of blood to epiphysis of head.

The *synovial membrane,* in addition to lining the capsule covers the neck up to the articular cartilage of the head, and dips in under the transverse ligament to cover the pad of fat and to ensheath the ligament of the head.

Nerve supply from the obturator, femoral, nerve to quadratus femoris and superior gluteal.

Arterial supply from the obturator, medial circumflex and gluteal. Head of femur in adult supplied with blood by arteries that pass along neck beneath the retinacular fibres of capsule; fracture of neck within the capsule thus likely to cause avascular necrosis of head of femur.

Note that the epiphysis of the head of the femur is entirely intracapsular.

Movements: Since the hip joint is a ball and socket joint flexion, extension, abduction, adduction and lateral and medial rotation are possible. In flexion and extension the head of the femur spins in the acetabulum round a transverse axis. Flexion, due to the iliopsoas and a number of other muscles (rectus femoris, pectineus, sartorius and tensor fasciae latae), amounts to about 50° if the knee is straight. If the knee is bent, flexion to about 90–100° is possible, due to the relaxation of the hamstrings. Extension is limited to about 15° due to the powerful ligaments on the anterior aspect of the joint. The main extensors are the hamstrings and gluteus maximus. Note that extension from the flexed position is a common movement (standing up, walking up stairs or up a slope).

Abduction in which the head of the femur slides and rotates about an anteroposterior axis amounts to about 50–60° and is due mainly to the gluteus medius and minimus. Adduction is limited by the opposite lower limb in the anatomical position. If the opposite limb is moved away adduction amounts to about 30° and is due to the adductors magnus, longus and brevis and gracilis.

Lateral and medial rotation (sliding and rotatory movements about a vertical axis through the head and intercondylar notch of the femur so that the front of the thigh faces laterally or medially) amounts to about 90°. Lateral rotation (about 60°) is due to the piriformis, obturators internus and externus, quadratus femoris and gluteus maximus. Medial rotation (about 30°) is due to the gluteus minimus and tensor fasciae latae.

When determining the range of flexion and extension movements at the hip joint it is important to fix the pelvis so that movements at the lumbar vertebrae do not add to the apparent range of movement. In walking the trunk moves on the lower limb when the foot is on the ground. For example, the trunk tilts on to the supporting limb, a

movement equivalent to abduction at the hip involving the
gluteus medius.

The knee joint (Figs. 14 and 15)

This is a synovial condyloid joint, but allows some rotation
in flexion. The expanded condyles of the lower end of the

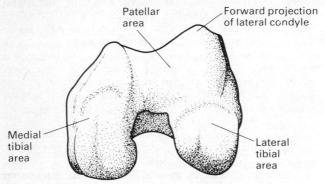

Fig. 14 Lower end of left femur to show differences between
condyles.

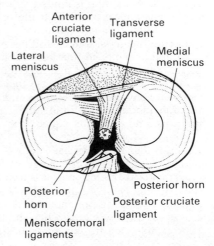

Fig. 15 Plateau of left
tibia from above.

femur articulate with the expanded upper end of the tibia, and also with the patella. As it is a weight-bearing joint it requires to be very stable; the expansion of the bony ends helps this, but there is no great adaptation of their shapes to one another; stability depends on very strong ligaments intra- and extra-articular, and on muscles.

Articulating surfaces: The lower end of the femur (Fig. 14); consists of the condyles for the tibia and the patella.

The femoral condyles project backwards and are separated by the intercondylar notch; the articular areas for the tibia are separated from the patellar articular area by shallow grooves. The medial condyle is lower (with the femur vertical), narrower and longer from before backwards than the lateral, and also curves outward from before backwards, whereas the lateral is straight. This is related to the lateral rotation of the tibia at the end of extension. Laterally the patellar surface is convex and prominent and medially concave. The posterior surface of the patella is concave laterally and convex medially. The lateral condyle of the tibia is broad from side to side, and the medial is oval with its long axis anteroposterior. These correspond with the femoral condyles. The posterior border of the lateral condyle of the tibia is rounded for the lateral meniscus (p. 35).

Capsule and ligaments: (1) The *capsule* is thin and much strengthened by special ligaments. Posteriorly and laterally it is attached above to the margins of the femoral condyles and the intercondylar line, and below to the margins of the tibial condyles; at the back of the lateral tibial condyle it is separated from the bone by the tendon of popliteus (p. 35).

Anteriorly the capsule is perforated above where the joint communicates with the suprapatellar bursa. The femoral epiphyseal line is within the capsule only anteriorly; the tibial epiphyseal line is entirely extracapsular.

(2) The *medial ligament* is a broad, flat, triangular band attached above to the medial epicondyle of the femur. Posterior fibres are attached to the medial meniscus; superficial fibres pass downwards and forwards to the medial

surface of the tibia and are attached for 10 cm below the condyle.

(3) The *lateral ligament*, round and cord-like, is quite free from the capsule. It is attached above to the lateral epicondyle of the femur, and below to the head of the fibula in front of its apex under cover of the biceps femoris tendon; its fibres run downwards and backwards.

(4) The *arcuate ligament* is a thin band arching from the head of the fibula across the popliteus to the capsule and the lateral meniscus.

(5) The *ligamentum patellae and retinacula:* The ligamentum patellae, continuous beyond the patella with the quadriceps tendon (p. 97), is thick and strong. It extends from the apex and lower border of the patella to the tubercle of the tibia. The retinacula are expanded lateral parts of the quadriceps tendon going to the tibia and reinforcing the ligamentum patellae. The quadriceps tendon, retinacula, patella and patellar ligament replace the capsule anteriorly.

Note: Between the ligamentum patellae and the depressed area of the tibia above the tuberosity are a pad of fat and a bursa; the fat adapts itself to fill in the varying open angle between the tibia and the femur.

(6) The *oblique posterior ligament* is an expansion upwards and laterally from the tendon of the semimembranosus to the back of the capsule. It is very strong and is perforated by the middle genicular vessels and nerve.

(7) The *cruciate ligaments* (Fig. 15) connect the femur to the tibia and lie within the fibrous capsule but outside the synovial membrane which ensheaths them anteriorly and at the sides. They cross each other with the anterior passing backwards lateral to the posterior.

The *anterior cruciate ligament* connects the front of the intercondylar fossa of the tibia to the back of the medial aspect of the lateral condyle. The *posterior cruciate ligament* passes forwards and connects the back of the intercondylar fossa of the tibia to the front of the lateral aspect of the medial condyle.

The menisci (semilunar cartilages): Each meniscus is a C-shaped fibrocartilaginous structure, wedge-shaped on transverse section, and avascular except at its attachment.

Base of wedge is attached to joint capsule except where popliteus tendon intervenes posteriorly behind the lateral meniscus. Thin edge of wedge insinuates between femur and tibia. Free ends (*horns* or *cornua*) of each meniscus are attached to plateau of tibia by fibrous tissue; anterior ends of menisci connected by transverse ligament. Posterior convexity of lateral meniscus is attached to medial condyle of femur by fibrous slips that embrace the posterior cruciate ligament (*ligament of Humphry* in front, *ligament of Wrisberg* behind). These are called *anterior* and *posterior meniscofemoral ligaments*. Upper portion of popliteus muscle is inserted into posterior convexity of lateral meniscus. During rotation of flexed knee, lateral meniscus held by meniscofemoral ligaments and by popliteus in position relative to femoral condyle. Medial meniscus, lacking such control mechanism, more mobile and liable to injury.

Synovial membrane: This lines the capsule, but leaves the capsule posteriorly to pass forwards round the cruciate ligaments which lie between it and the back of the capsule. Above the patella it communicates with suprapatellar bursa extending about 6 cm up the shaft of the femur. Below the patella the synovial membrane is pushed into the joint by the pad of fat, and forms a fold or ridge on each side of the pad, the *alar folds*. In the midline a thin double layer of membrane is attached to the anterior edge of the intercondylar notch of the femur, the *infrapatellar fold (ligamentum mucosum)*. The menisci are not covered by synovial membrane.

Bursae communicating with knee-joint: (1) the *suprapatellar bursa*; (2) the *popliteus bursa*, which is an extension of the synovial membrane round the popliteus as it emerges from the capsule and often opens into superior tibiofibular joint; (3) frequently the bursa between the medial head of gastrocnemius and the medial condyle.

Bursae in neighbourhood not opening into joint: (1) *prepatellar*, subcutaneous over patella, swollen in *housemaid's knee*; (2) *infrapatellar*, (a) subcutaneous over ligamentum patellae, swollen in *clergyman's knee*, and (b) deep to ligamentum patellae in front of tibia; (3) between insertions of sartorius, gracilis and semitendinosus; (4)

beneath semimembranosus tendon, often communicating with medial gastrocnemius bursa and thence with joint; (5) under lateral head of gastrocnemius; (6) between lateral ligament and biceps femoris tendon.

Nerve supply from femoral, obturator and sciatic nerves.

Blood supply: genicular branches of popliteal artery, middle genicular direct to cruciate ligaments.

Movements: Flexion and extension, and lateral and medial rotation (about total of 40°) when flexed. Flexion is backward movement of the leg on the thigh or vice versa. Extension is restoration of the flexed leg or thigh to the anatomical position. At the end of extension of the leg on the thigh there is some lateral rotation of the tibia on the femur and at the end of extension of the thigh on the leg there is some medial rotation of the femur on the tibia. The adaptation of the bony surfaces to this has been described (p. 33); the cause of the rotation is the pull of the quadriceps extensor against the tight oblique and anterior cruciate ligament. The completely extended knee is said to be locked because it will maintain the body in the upright position without muscular effort. The weight of the body passes just in front of the centre of the knee, and in front of the medial and lateral ligaments, so that these and the anterior cruciate and posterior ligaments resist the weight and enable the muscles to relax. Flexion from this position is initiated by the popliteus, which rotates the femur laterally or the tibia medially, and is said to unlock the joint.

Abduction and adduction of the leg on the thigh is prevented by the medial and lateral ligaments respectively. The rotatory movements are limited by the same ligaments as well as the cruciate ligaments. Sliding of the tibia on the femur or vice versa in a horizontal plane is prevented by the cruciate ligaments.

The tibiofibular joints

The fibula is very strongly bound to the tibia below, but has a small synovial joint at its upper end.

Superior tibiofibular joint has a plane oval facet on the posterolateral angle of the lateral condyle of the tibia (on the epiphysis) which faces downwards and laterally to articulate with a similar facet on the head of the fibula. The joint has a capsule with anterior and posterior ligaments. Its nerve supply is from nerve to popliteus and recurrent genicular nerve.

The interosseous membrane and tibia and fibula constitute the *middle tibiofibular joint*. The fibres of the membrane are mostly directed downwards and laterally. It is perforated above by the anterior tibial vessels which pass forwards, and below by the peroneal artery.

The *inferior tibiofibular joint (syndesmosis)* consists of a thick *interosseous ligament* joining rough triangular areas of more than 3 cm in vertical extent on the two bones. There are in addition *anterior* and *posterior inferior tibiofibular ligaments*; the posterior extends below the lower edge of the tibia and forms part of the socket receiving the talus.

There is an extension of the ankle joint into the lowest part of the tibiofibular joint where the surfaces are covered with articular cartilage.

Fibular facet of talus is commonly convex in anteroposterior direction; hence dorsiflexion at ankle joint rotates fibula laterally about 5° about its long axis, with movement at superior synovial joint.

The strength of the inferior joint is a vital factor in the stability of the ankle joint. Obviously the weight of the body and the counter-thrust of the talus would separate the two bones were they not very strongly bound together.

The ankle (talocrural) joint (Figs. 16 and 17)

This is a synovial hinge joint with good adaptation of the articular surfaces, but its strength also depends on its ligaments and surrounding muscles.

Articulating surfaces: The lower ends of the tibia and fibula (strongly bound together, p. 37) form a deep mortice to receive the body of the talus. Only the tibia articulates with the upper surface of the talus; the malleoli of the tibia and fibula grip the talus at the sides. The facet on the medial side of the talus is comma-shaped, with a large non-articular area below it; the lateral facet is triangular (apex down), and covers the whole of the side of the body of the

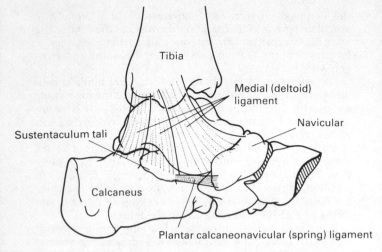

Fig. 16 Medial side of left ankle joint to show attachments of deltoid ligament.

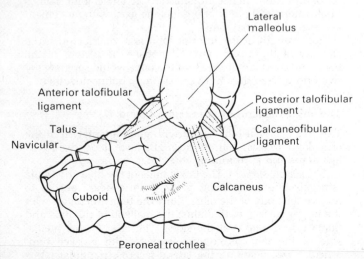

Fig. 17 Lateral side of left ankle joint.

talus. Reciprocally, the lateral malleolus extends further distally (1 cm) than the medial, and strongly resists eversion at the ankle. The socket and the body of the talus are wider in front than behind, so that the dorsiflexed foot is locked, and the plantar flexed foot can move slightly from side to side.

Note that the socket is deepened posteriorly by the inferior tibiofibular ligament, which has a special facet along the lateral edge of the body of the talus.

Capsule and ligaments: The capsule is thin, and attached round the margins of the articular surfaces, except anteriorly where it is attached to the neck of the talus distal to the articular edge.

The *deltoid (medial collateral) ligament* is strong and triangular (Fig. 16). Deep fibres run from the medial malleolus to the rough surface on the medial aspect of the body of the talus. Superficial fibres spread out from the malleolus to the tuberosity of the navicular, plantar calcaneonavicular (spring) ligament (p. 41), sustentaculum tali and medial tubercle of the talus.

The *lateral ligament* consists of three distinct bands: (1) *anterior talofibular ligament,* between front of malleolus and lateral side of neck of talus; (2) *calcaneofibular ligament* between malleolus and a tubercle on the lateral surface of the calcaneus above and behind the peroneal tubercle; (3) *posterior talofibular ligament,* from the deep pit behind articular surface of lateral malleolus (malleolar fossa) horizontally to lateral tubercle of talus; it produces a groove on the talus.

The *synovial membrane* lines the capsule and extends (a) on to the neck of the talus in front of the joint, and (b) upwards between the lower ends of the tibia and fibula (p. 37).

Arteries and *nerves* enter the joint from malleolar arteries and from anterior and posterior tibial nerves.

Movements: There are about 60° of downward movement of the foot, plantar flexion (flexion) and about 45° of dorsiflexion (extension) upward movement of the foot. When the toes are pointed downwards there is slight abduction and adduction, owing to the posterior narrow part of the talus lying in the wider socket. The main plantar flexors are the soleus and gastrocnemius. They may be assisted by the long flexor muscles entering the sole. The main dorsiflexors are tibialis anterior and the long extensors of the toes passing over the dorsum of the foot. Note the action of the large calf muscles in raising the body on to the toes when the foot is on the ground.

The tarsal joints (Fig. 18)

These are mainly gliding synovial joints allowing very little movement. Important movements, however, take place between the talus and the rest of the foot as a whole. These

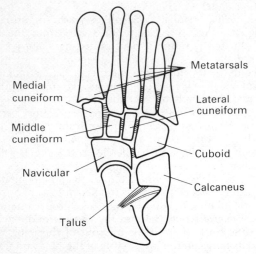

Fig. 18 Horizontal section through foot to show bones and interosseous ligaments.

occur at what is clinically called the *subtalar joint*, which is a composite of two joints—the posterior talocalcaneal and the talocalcaneonavicular.

The posterior talocalcaneal joint
Between the body of the talus and the body of the calcaneus (often called the subtalar joint).

Articulating surfaces: A concave facet on the posterior part of the inferior surface of the body of the talus and a corresponding convex facet on the middle part of the superior surface of the calcaneus.

Capsule and ligaments: A capsule surrounds the joint. The anterior part is really the posterior fibres of the interosseous ligament.

Interosseous talocalcaneal ligament consists of two strong bands lying obliquely in the sinus tarsi formed by the sulcus calcanei and sulcus tali and a third more laterally in the sinus tarsi connecting calcaneus to neck of talus (cervical ligament).

Movements are described below, since this joint and the next function as one.

The talocalcaneonavicular joint

Articulating surfaces: The convex head of the talus and a concave socket formed by the calcaneus, navicular and ligaments. (1) Anteriorly the head fits into the hollow posterior surface of the navicular. (2) Inferiorly head of talus rests on facets on the upper surface of the sustentaculum tali and anterior part of calcaneus. These two facets may fuse into one. (3) In front of the sustentaculum tali the head of the talus rests on the plantar calcaneonavicular (spring) ligament. (4) Laterally and inferiorly the socket is completed by the medial (calcaneonavicular) limb of the bifurcate ligament.

Note: The spring ligament extends from the front of the sustentaculum tali to the tuberosity of the navicular and the under surface of that bone. It is strong and is an important support of the head of the talus in the medial arch of the foot. The bifurcate ligament is a Y-shaped band extending from the anterior dorsal part of the calcaneus anteriorly to the navicular (medial limb) and cuboid (lateral limb).

Movements: Movements at these two joints occur at the same time. The foot is inverted and everted by approximately ball-and-socket movements round the head of the talus at the talocalcaneonavicular joint, associated with gliding movements in the subtalar joint. The fulcrum of these movements is the strong interosseous ligament.

Inversion, in which the sole of the foot faces inwards and its inner border moves inwards and upwards, is produced by simultaneous contraction of tibialis anterior and tibialis posterior.

Eversion, in which the sole of the foot faces outwards and the lateral border moves outwards and upwards, is produced by the peronei (longus, brevis and tertius). The movement begins in the midtarsal joint (calcaneocuboid and talonavicular) but is restricted here. Continuing movement goes on in the subtalar joints. Axis of this latter movement passes through centres of curvature of the two subtalar joints; it is oblique from lateral tubercle of calcaneus upwards and medially through neck of talus. Thus inversion is accompanied by adduction of front of foot and eversion by abduction. Inversion is most marked with the foot plantar flexed and eversion with the foot dorsiflexed. This is due to the arrangement of the medial and lateral ligaments of the ankle joint.

The other tarsal joints do not require special descriptions, but the following features should be noted:

The *plantar calcaneocuboid (short plantar) ligament* is attached to a groove at the anterior end of the plantar surface of the calcaneus and to the ridge forming the posterior boundary of the groove on the cuboid.

The *long plantar ligament* lies superficial to the short. It extends from the anterior margins of the tubercles of the calcaneus to the cuboid, covers over the peroneus longus tendon in the groove, and reaches the plantar surfaces of the lateral cuneiform and the bases of the 2nd, 3rd and 4th metatarsals. It is an important support of the lateral arch of the foot.

Interosseous ligaments join the three cuneiforms to one another, the lateral cuneiform to the cuboid, and the navicular to the cuboid.

Tarsometatarsal joints

The 1st metatarsal has a large kidney-shaped facet for the front of the medial cuneiform; the 4th and 5th metatarsals articulate with the cuboid. The first joint has a definite capsule, the others have dorsal and plantar ligaments, and the bases of all the metatarsals are joined by interosseous ligaments.

Note: The line of these joints across the foot is not transverse, nor is it even. The base of the 2nd metatarsal dips into a mortice between the medial and lateral cuneiforms, to reach the short intermediate cuneiform. Note also that the medial side of the base of the 2nd metatarsal is connected to the 1st cuneiform by a strong interosseous ligament, so that the big toe has almost no mobility at its tarsometatarsal joint. Compare this with the mobility at the carpometacarpal joint of the thumb. The loss of opposition by the big toe is a striking characteristic of the human foot.

The metatarsophalangeal and interphalangeal joints

These joints resemble those of the hand and do not require further description. Note, however, that the heads of the metatarsals are more tightly held together than those of the metacarpals.

Plantar flexion and dorsiflexion take place at all these joints. Dorsiflexion at metatarsophalangeal joints is much greater than at interphalangeal and important in the late stance phase in walking. Easily demonstrated by standing on one's toes.

Some abduction and adduction occur at the metatarsophalangeal joints with reference to a long axis through the second toe (dorsal interossei are abductors, plantar interossei are adductors).

Arches of the foot

Longitudinal

Medial: Consisting of calcaneus, talus, navicular, three cuneiforms and medial three metatarsals.

Lateral: Calcaneus, cuboid, lateral two metatarsals.

Transverse

Bases of metatarsals: Each foot really half a dome, lateral side on ground, medial side at upper limit of dome.

Factors maintaining the arches

Medial longitudinal. (1) *Muscular*—flexor hallucis longus lies below arch and gives slip to 2nd and 3rd toes; flexor digitorum longus; tibiales anterior and posterior; flexor digitorum brevis-medial part, abductor hallucis. (2) *Ligamentous*—plantar aponeurosis, spring ligament, interosseous ligaments.

Lateral longitudinal. (1) *Muscular*—peroneus longus (like a sling) flexor digitorum longus to 4th and 5th toes; flexor digitorum brevis-lateral part, abductor digiti minimi. (2) *Ligamentous*—plantar aponeurosis; long and short plantar ligaments, interosseous ligaments.

Transverse. (1) *Muscular*—peroneus longus; adductor hallucis. (2) *Ligamentous*—interosseous litaments. (3) *Bony*—wedge shape of intermediate and lateral cuneiforms and bases of middle three metatarsals (base of wedge is dorsal).

Arches give resilience to the foot but although segmentation permits mobility it also makes the arches unstable.

2
The Muscles

INTRODUCTION

The terms *origin* (the fixed part of the muscle) and *insertion* (the part that moves) are retained because they are concise, but it cannot be too strongly urged that the student should appreciate that the reverse can occur and also that both ends may move. When a muscle contracts it shortens. Which end moves and which remains stationary depends on circumstances.

The muscles which produce a movement are called *prime movers*. Muscles which have to lengthen or relax to allow a movement are called *antagonists*. *Synergists* are muscles which contract to prevent an unwanted movement by the prime movers, and *fixators* contract to prevent movement at a joint not acted on by the prime movers. In many movements gravity is the main force, and a muscle opposing gravity controls the movement so that a muscle is involved in an action which is the opposite of that usually ascribed to it, e.g. in sitting down from the standing position the quadriceps femoris controls the flexion at the knee. This is sometimes called the *action of paradox*.

The term *raphe* is used to denote the interdigitation of musculotendinous fibres. A *ligament* or *tendon* consists of longitudinal fibres which cannot be elongated unless it is slack and the slack is taken up. An *aponeurosis* is a large flat tendon.

THE MUSCLES OF THE HEAD AND NECK

THE MUSCLES OF FACIAL EXPRESSION

The musculature of the 2nd pharyngeal (branchial) arch during development forms a thin sheet which spreads over the neck (*platysma*), face (muscles of expression) and scalp (muscles of scalp and of external ear); these are all supplied by the facial (7th cranial) nerve. Some of the muscles of the

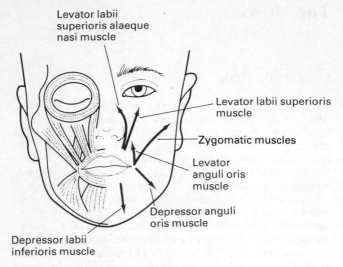

Fig. 19 Principal muscles of face.

face are arranged around the apertures (eye, nose and mouth). Each aperture possesses a sphincter and a dilator mechanism.

Platysma
Its origin is from the fascia over the upper part of the chest and shoulder. Forms a wide sheet covering side of neck with the fibres directed upwards and medially. Its insertion is into lower border of mandible and spreads up on to face as risorius attached to angle of mouth.

Note: The platysma lies in the deep part of the superficial fascia of the neck; it covers the external jugular vein and the cutaneous branches of the cervical plexus (p. 269).

Muscles of the eyelids

Orbicularis oculi is the sphincter muscle and consists of three parts:
(1) *Orbital part* (surrounding orbit)

Origin: Frontal and maxillary bones at medial orbital margin.

Insertion: Skin only; encircles orbit and has no bony insertion laterally.

(2) *Palpebral part* (slightly arched fibres in eyelids)

Origin: Medial palpebral ligament.

Insertion: Lateral palpebral raphe.
Note: Palpebral ligament and raphe are attached to medial and lateral bony margins of orbit, and split to go to edges of eyelids.

(3) *Lacrimal part* attached to lacrimal sac

Origin: Lacrimal crest, forming posterior edge of lacrimal fossa.

Insertion: Deep aspect of medial palpebral ligament.

Nerve supply of the whole muscle is facial nerve.

Actions: Very different in the two parts.

The palpebral part closes the lids gently edge to edge, with no diminution in volume of conjunctival sac, so that tears are not extruded, but on the contrary by blinking are pumped medially via lacrimal sac into the nasal cavity. This part is opposed by levator palpebrae superioris.

The orbital part closes the lids forcibly, with diminution in volume of the conjunctival sac. If the sac is brimful of tears they spill over down the cheeks. Opponent of this part is occipitofrontalis.

The *levator palpebrae superioris* arises within the orbit and is inserted into the skin, tarsal plate and conjunctiva of the upper lid. It is antagonist to the orbicularis oculi and is supplied by the oculomotor (3rd cranial) nerve. Inferior part of levator is smooth muscle, supplied by sympathetic from superior cervical ganglion via carotid plexus. Upper lid droops, *ptosis*, if sympathetic pathway is interrupted.

The muscles of the eyeball are described with the visual apparatus (p. 323).

Corrugator supercilii
This muscle acts on the skin of the forehead, but is described as a muscle of the eyelids, being a detached portion of the orbicularis oculi.

Origin: Frontal bone at medial end of superciliary ridge.

Insertion: Fibres directed upwards and laterally through orbicularis oculi to skin of forehead.

Action: Produces frown of vertical wrinkles above nose.

Muscles of the scalp and external ear

Note that these belong to the same group as the muscles of expression, and are supplied by the facial nerve.
The scalp consists of three layers closely bound together: (1) skin, (2) superficial fascia, and (3) strong deep fascia (*epicranial aponeurosis*) with occipitalis muscle attached to it behind and frontalis muscle in front. These three layers move as a whole on the periosteum of the skull (*pericranium*), and are separated from this by a space containing loose connective tissue.

Occipitalis
Origin: Lateral two-thirds of highest nuchal line.

Insertion: Posterior part of epicranial aponeurosis.

Action: Pulls scalp backwards, or fixes scalp to allow frontalis to pull on skin of forehead. It is supplied by the posterior auricular nerve from facial nerve.

Frontalis
Origin: Anterior part of epicranial aponeurosis.

Insertion: Skin of forehead.

Action: Produces broad wrinkles across forehead. The

muscle elevates the eyebrows (as in looking upwards) and is the opponent of the orbital part of orbicularis oculi. The scalp may be regarded as a prolongation of the facial skin and muscles over the skull vertex to the highest nuchal line.

Auricularis muscles
These are three small muscles (*anterior*, *superior* and *posterior*) inserted into the cranial surface of the pinna.

Muscles of the nose

Compressor naris (the sphincter)
 Origin: Maxilla at side of lower part of bony anterior nares.

 Insertion: Aponeurosis continued across nose to fellow of opposite side.

 Action: Compresses upper part of cartilaginous nasal aperture.

Dilatator naris
 Origin: Maxilla below compressor naris.

 Insertion: Side of alar cartilage.

 Action: Pulls on lateral wing of cartilage to open up nostril.

These two muscles constitute the *nasalis* muscle and the compressor is called the transverse part and the dilatator the alar part.

Procerus
A continuation of median fibres of frontalis down on to nose.

 Origin: Fascia over bridge of nose.

 Insertion: Skin of lower and median part of forehead.

Action: Produces small transverse wrinkles at root of nose by slight elevation of external nose.

Muscles of the mouth

Sphincter of mouth

Orbicularis oris (Figs 19 and 20): Has its own proper fibres attached to bone above and below in midline, the *superior* and *inferior incisive slips*. Its bulk is much increased by fibres

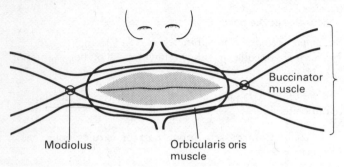

Fig. 20 Arrangement of fibres in orbicularis oris muscle.

received from buccinator (p. 51) and the dilatator muscles described below. It encircles the lips.

Action is to close the mouth into a small circle, as in whistling.

Dilatators of mouth

Are arranged radially around the lips; consist of elevators of upper lip and angle of mouth and depressors of lower lip and angle of mouth.

Upper lip: The *levator labii superioris* arises from the infra-orbital margin; inserted into the upper lip.

The *levator labii superioris alaeque nasi* arises from the upper part of frontal process of maxilla; the labial part blends with levator labii superioris, and the nasal part is inserted into alar cartilage of nose.

Labial part elevates lip, nasal part dilates nostril.

Angle of mouth (elevators): Zygomaticus major and *zygomaticus minor* arise from zygomatic bone.

Levator anguli oris arises from canine fossa of maxilla (it lies under cover of levator labii superioris and infra-orbital nerve).

These three muscles descend to be inserted into the angle of the mouth at the modiolus.

Lower lip: The *depressor labii inferioris* arises from the front of the mandible near the midline, and is inserted into the lower lip.

Angle of mouth (depressor): Depressor anguli oris arises from anterior part of external oblique line of mandible and is inserted into angle of mouth at modiolus.

The *buccinator* (Figs 20 and 21) forms the muscular plane of the cheek. It has a deep posterior origin from the pterygo-mandibular raphe which is attached above to the hamulus

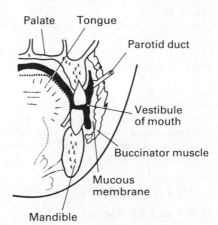

Fig. 21 Coronal section through jaws and mouth.

of the medial pterygoid plate and below to the posterior end of the mylohyoid line of the mandible. The buccinator also arises from the adjacent areas of the outer surfaces of the maxilla and mandible near the three molar teeth. It is inserted into the orbicularis oris—upper and lower fibres into respective lips, central fibres decussate at modiolus.

Its action is to obliterate the space between cheek and teeth. It is used in chewing and sucking and pushes the food between the teeth.

Note: The *pterygomandibular raphe* is attached above to the hamulus of the medial pterygoid plate and below to bone beside posterior border of third molar tooth at the posterior end of the mylohyoid line of the mandible. It unites the buccinator and the superior constrictor (p. 336), these muscles being in the same plane.

The buccinator is pierced by the parotid duct, buccal branch of mandibular nerve and ducts of molar glands. Tendon of tensor veli palatini passes above its border between tuberosity of maxilla and hamulus.

The angles of the mouth also receive the *risorius*, a prolongation on to the face of some fibres of the platysma (p. 46).

The modiolus
The point of decussation of central fibres of buccinator also receives elevators and depressors of angle of mouth. Point of crossing of so many fibres makes palpable nodule inside angle of mouth opposite first upper premolar tooth

Facial (Bell's) palsy results from the facial nerve not functioning. The face is pulled to the unaffected side and the patient cannot wrinkle the forehead, frown, close the eye, twitch the nose, smile or whistle on the affected side. The patient complains of food collecting between the cheek and gum.

OTHER MUSCLES OF THE HEAD

Muscles of mastication (Fig. 22)

These are powerful muscles acting on the mandible and affecting the movements of mastication. They are all developed from the 1st pharyngeal (mandibular) arch and supplied by the mandibular division of the trigeminal (5th cranial) nerve (p. 256).

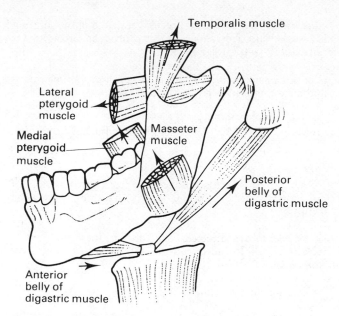

Fig. 22 Actions of muscles of mastication.

Masseter
Origin: Anterior two-thirds of lower border of zygomatic arch (superficial head): posterior one-third of lower border and whole of deep surface of arch (deep head).

Insertion: Outer surface of ramus of mandible from mandibular notch to angle.

Nerve supply: Masseteric nerve (mandibular division of trigeminal nerve) through mandibular notch.

Action: Closes the jaw.

Temporalis
Origin: From the inferior temporal line and the temporal fossa and fascia.

Insertion: The coronoid process and a hollow running downwards on deep aspect of this process as far as the last molar tooth (retromolar fossa).

Nerve supply: Two or three deep temporal nerves (mandibular nerve) which enter its deep aspect.

Action: Closes the jaw.
Note: The most posterior fibres are horizontal and retract the jaw.

Lateral pterygoid
  Origin: By two heads—lower head from lateral surface of lateral pterygoid plate, upper head from infratemporal crest and fossa.

Insertion: Fibres directed backwards to fossa on front of neck of mandible and to capsule and disc of temporomandibular joint.

Nerve supply: From mandibular nerve.

Action: Involved in opening mouth by pulling condyle forwards as it rotates, and also protrudes jaw. One muscle acting alone pulls chin over to opposite side.

Medial pterygoid
  Origin: From medial surface of lateral pterygoid plate and from the tuberosity of the maxilla (lateral to the lateral pterygoid muscle).

Insertion: Fibres pass downwards, backwards and laterally to medial surface of ramus of mandible, from groove for mylohyoid nerve to angle.

Action: Closes the jaw. One muscle acting alone helps to pull chin over to opposite side.

Nerve supply: From trunk of mandibular nerve near otic ganglion.

MUSCLES OF THE NECK

Superficial muscles of the side of the neck

The *platysma* is described with the muscles of facial expression (p. 46) and the *trapezius* is described with the muscles of the upper limb (p. 70).

Sternocleidomastoid (Fig. 23)

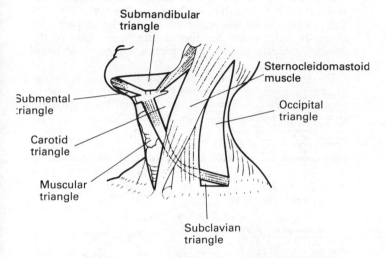

Fig. 23 Triangles of neck and muscles bounding them.

Origin: Sternal head: From front of manubrium sterni by a rounded tendon. *Clavicular head:* From upper border and front of medial one-third of clavicle by muscular fibres.

Insertion: Outer surface of mastoid process and lateral two-thirds of superior nuchal line. Medial fibres from clavicle pass upwards deep to accessory nerve and to sternal belly of the muscle to the mastoid process and form the cleidomastoid part of the muscle.

Nerve supply: Motor—spinal part of accessory (11th cranial) nerve; sensory (proprioceptive)—ventral rami of 2nd and 3rd cervical nerves.

Action: The two muscles together flex the head and neck. One alone pulls head down towards shoulder of same side and turns face to opposite side and elevates the chin.

Note: Torticollis is due partly to contraction of this muscle, and the position of the head illustrates the action of one muscle.

Important relations of sternocleidomastoid: It separates the anterior from the posterior triangle of the neck (Fig. 23). The carotid sheath lies under cover of its lower part, and along its anterior border above. The upper part of the muscle conceals the cervical plexus lying on scalenus medius.

The nerves of the cervical and brachial plexus appear at its posterior border. The spinal part of the accessory nerve runs backwards and downwards through its deep fibres and emerges halfway down its posterior border.

Cervical fascia (Fig. 24)
This consists of an *investing (superficial) layer* around the neck and a complex system of septa between the neck

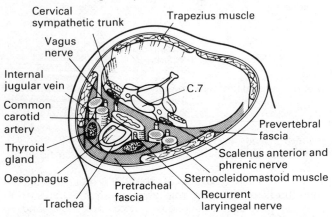

Fig. 24 Diagram of compartments into which neck is divided by deep cervical fascia.

muscles and other structures. The investing layer is attached behind to the ligamentum nuchae and is continuous round the neck. It splits to ensheath the trapezius and sternocleidomastoid muscles. It is attached below to the clavicle and to the upper border of the sternum; the sternal attachment splits into anterior and posterior layers which enclose the *suprasternal space* containing the anterior jugular veins. The clavicular part is likewise split, its deep lamina binding the inferior belly of omohyoid to the clavicle. External jugular vein pierces both laminae below omohyoid, just above midpoint of clavicle.

Above, it is attached to the lower border of the mandible, splits to enclose the submandibular gland, covers the masseter and encloses the parotid gland. It is attached to the zygomatic arch, supramastoid crest and lateral third of the superior nuchal line.

A transverse septum passing in front of the prevertebral muscles (p. 61) and behind the pharynx and oesophagus constitutes the *prevertebral fascia*; this fascia extends laterally and covers cervical and brachial plexuses and muscles in floor of posterior triangle. Attached above to the base of the skull, it extends downwards into superior mediastinum, attached to body of 4th thoracic vertebra. It forms a base upon which pharynx, oesophagus and carotid sheath slide in swallowing and neck movements.

The *carotid sheath* is areolar tissue, strong over carotid arteries, weak over jugular vein (to allow for expansion of latter). Is attached above near tympanic plate of temporal bone.

The *pretracheal fascia* is attached to the thyroid and cricoid cartilages. It invests the thyroid gland and passes down into the thorax behind the brachiocephalic veins to the pericardium. Laterally it blends with front of carotid sheaths.

The suprahyoid muscles

Digastric consists of two bellies connected by an intermediate tendon.

Origin: Anterior belly from digastric fossa on back of body of mandible near the midline; *posterior belly* from mastoid notch on base of skull under cover of mastoid process.

Insertion: Each belly is inserted into intermediate tendon, which is not attached to bone but held in a loop of fascia which holds it down to the side of the body and adjacent part of greater cornu of hyoid bone.

Nerve supply: Anterior belly, mylohyoid branch of inferior alveolar nerve, a branch of the mandibular; posterior belly, facial nerve near exit from stylomastoid foramen (derivatives of 1st and 2nd pharyngeal arches respectively).

Action: Depression of mandible or elevation of hyoid bone.
Note: The digastric bounds the digastric triangle (Fig. 23) containing the submandibular gland. The muscle is a landmark for the facial (7th cranial) nerve (above it) and the hypoglossal (12th cranial) nerve (below it), while the occipital artery runs under cover of its posterior belly.

Stylohyoid
A small slip lying along the upper border of the posterior belly of the digastric, derived like it from the 2nd pharyngeal arch, and supplied by the facial nerve.

Origin: Styloid process.

Insertion: Fibres split over digastric tendon to reach body and lesser cornu of hyoid bone.

Action: Elevation of hyoid bone in swallowing.

Mylohyoid
A flat thin sheet of muscle. The mylohyoids form the floor of the mouth.

Origin: Whole length of mylohyoid line of mandible.

Insertion: Fibres are directed medially to meet muscle of opposite side in a median raphe which extends from the mandible to the hyoid bone; the most posterior fibres are inserted into the body of the hyoid bone.

Nerve supply: Mylohyoid branch of inferior alveolar nerve, a branch of the mandibular on its inferior surface (cf. anterior belly of digastric).

Action: Elevation of hyoid and floor of mouth. Presses up tongue in swallowing. If hyoid bone is fixed can depress mandible.

Geniohyoid
Really an upward extension of pretracheal infrahyoid muscles, and is in series with rectus abdominis. All are supplied segmentally by spinal nerves.

Origin: Inferior genial tubercle on back of symphysis menti.

Insertion: Anterior surface of body of hyoid bone above mylohyoid.

Nerve supply: 1st cervical spinal nerve (through hypoglossal).

Action: If hyoid bone is fixed, can depress mandible.

Muscles of tongue are described in the section on the tongue (p. 327).
Muscles of pharynx are described in the section on the pharynx (p. 336).
Muscles of palate are described in the section on the palate (p. 331).
Muscles of larynx are described in the section on the larynx (p. 370).

The infrahyoid muscles (Fig. 25)

Sternohyoid
Narrow strap-like muscle converging upwards in front of neck.

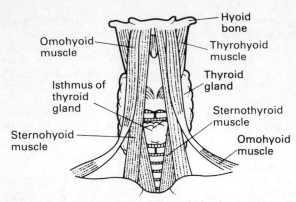

Fig. 25 Infrahyoid muscles and thyroid gland.

Origin: Back of sternoclavicular joint and adjacent parts of clavicle and manubrium.

Insertion: Lower border of body of hyoid bone close to midline.

Nerve supply: Ansa cervicalis (C. 1, 2, 3).

Action: Depression of hyoid bone.

Sternothyroid
Wider muscle, under cover of sternohyoids and diverging upwards.

Origin: Back of first costal cartilage and adjacent part of manubrium sterni.

Insertion: Oblique line on outer aspect of thyroid cartilage.

Nerve supply: Ansa cervicalis (C. 1,2, 3).

Action: Depression of larynx.
Note: Diamond-shaped space in midline between the muscles of the two sides contains trachea, cricoid and thyroid cartilages, isthmus of thyroid gland and inferior thyroid veins.

Omohyoid
Two bellies united by an intermediate tendon; the muscle extends obliquely across the neck from scapula to hyoid bone.

Origin: Suprascapular ligament and adjacent part of upper border of scapula.

Insertion: Lower border of body of hyoid bone lateral to sternohyoid.
Note: Intermediate tendon lies over carotid sheath and under sternocleidomastoid, and is held down to clavicle by a fascial investment (Figs. 23 and 25).

Nerve supply: Superior belly from superior branch (descendens hypoglossi) of ansa cervicalis (1st cervical nerve); inferior belly from ansa cervicalis (1st, 2nd, 3rd cervical nerves).

Action: Pulls hyoid bone downwards and backwards.
Note: Anterior belly separates carotid triangle above from muscular triangle below. Posterior belly separates occipital triangle above from supraclavicular triangle below.

Thyrohyoid
Origin: Oblique line of ala of thyroid cartilage.

Insertion: Lower border of greater cornu of hyoid bone.

Nerve supply: 1st cervical nerve through hypoglossal.

All the infrahyoid muscles are used in chewing, swallowing and speech. They fix the hyoid bone so that the suprahyoid muscles have a fixed point from which to act.

The prevertebral muscles

These form a group lying in front of the cervical vertebrae.

Longus cervicis consists of three parts.

Vertical: Slips from front of bodies of upper four thoracic vertebrae to front of bodies of 2nd, 3rd and 4th cervical vertebrae.

Lower oblique: Slips from upper thoracic vertebrae directed upwards and laterally to transverse processes (anterior tubercles) of 5th and 6th cervical vertebrae.

Upper oblique: Slips from transverse processes (anterior tubercles) of 3rd, 4th and 5th cervical vertebrae directed upwards and medially and converging on to anterior tubercle of atlas.

Nerve supply: Ventral rami of 2nd, 3rd, 4th cervical nerves.

Action: All parts acting together flex the neck. Oblique parts on one side produce lateral flexion.

Longus capitis:
Origin: Like scalenus anterior from anterior tubercles of transverse processes of 3rd, 4th, 5th and 6th cervical vertebrae.

Insertion: Base of skull on inferior surface of basilar part of occipital bone.

Nerve supply: Ventral rami of 1st, 2nd, 3rd cervical nerves.

Action: Flexion of head and neck.

Rectus capitis anterior:
Origin: Front of lateral mass of atlas.

Insertion: Basilar part of occipital bone deep to longus capitis.

Nerve supply: Ventral rami of 1st, 2nd, 3rd cervical nerves.

Action: Flexion of head.

Rectus capitis lateralis:

Origin: Transverse process of atlas.

Insertion: Inferior surface of jugular process of occipital bone.

Nerve supply: Ventral ramus of 1st cervical nerve.

Action: Lateral flexion of head to same side.

These short muscles and the suboccipital muscles are responsible for the fine adjustment of the position of the head on the vertebral column.

Three muscles connect the cervical vertebrae to the 1st and 2nd ribs.

Scalenus anterior:

Origin: Anterior tubercles of transverse processes of 3rd, 4th, 5th and 6th cervical vertebrae.

Insertion: Scalene tubercle on medial border of 1st rib, by narrow tendon.

Nerve supply: Ventral rami of 4th, 5th cervical nerves.

Scalenus medius

Origin: Posterior tubercles of transverse processes of 2nd to 6th cervical vertebrae.

Insertion: Into large area of upper surface of 1st rib behind subclavian groove.

Nerve supply: Ventral rami of 3rd to 7th cervical nerves.
Note: Brachial plexus and subclavian artery emerge between these two muscles.

Scalenus posterior
A special part of the scalenus medius passing across the 1st rib to reach the 2nd rib.

Origin: Posterior tubercles of transverse processes of 5th, 6th and 7th cervical vertebrae.

Insertion: Posterior part of outer surface of 2nd rib.

Nerve supply: Ventral rami of 5th to 7th cervical nerves.

Actions: The scalene muscles of one side produce lateral flexion of the neck. Acting together the muscles of both sides flex the neck. With the neck fixed, they can be important muscles of inspiration since elevation of the 1st and 2nd ribs moves the whole rib cage.

THE MUSCLES OF THE DORSAL ASPECT OF THE TRUNK AND NECK

The trapezius, latissimus dorsi, levator scapulae and rhomboids are muscles of the upper limb (pp. 70–71).

Deep to these are the splenius muscles and the posterior serrati; and deep to these are the group of dorsal muscles of the vertebral column and the special suboccipital group attached to the axis, atlas and skull.

Splenius and posterior serrati muscles

Splenius capitis and *cervicis*
Origin: Lower half of ligamentum nuchae and spines of 7th cervical and first 5 or 6 thoracic vertebrae.

Insertion: Muscles form flat wide band directed upwards and laterally to mastoid process and occipital bone deep to sternocleidomastoid (capitis) and into posterior tubercles of transverse processes of 2nd and 3rd cervical vertebrae (cervicis).

Nerve supply: Dorsal rami of lower cervical nerves.

Serratus posterior superior
Origin: Spines of 7th cervical and upper three thoracic vertebrae deep to splenius.

Insertion: Thin sheet of fibres directed downwards and laterally to outer surfaces of 2nd to 5th ribs beyond angles.

Nerve supply: 3rd and 4th intercostal nerves (i.e. ventral rami of thoracic nerves).

Action: Elevator of upper ribs.

Serratus posterior inferior
Origin: Spines of last two thoracic and upper two lumbar vertebrae.

Insertion: Fibres directed laterally and upwards to outer surfaces of lower four ribs beyond angles.

Nerve supply: 9th, 10th and 11th intercostal nerves (i.e. ventral rami of thoracic nerves).

Action: Depresses or fixes lower ribs.

Deep muscles of vertebral column

The extensor muscles of the vertebral column include not only those inserted into the vertebrae, but certain longitudinal muscles inserted into the ribs near the vertebral column, and some partially inserted into the back of the head. All lie deep to the thoracolumbar fascia.

These muscles may conveniently be divided into three groups:

(1) *Short muscles between adjacent vertebrae*
These are *rotatores* thoracic region only (from transverse process upwards and medially to spine of vertebra above),

intertransverse (between adjacent transverse processes) and *interspinales* (between adjacent spines). They are the deepest of the series.

(2) *Muscles running obliquely from transverse processes below to spines or laminae above, missing several vertebrae between origin and insertion*

These include the *multifidus*, which miss one or two vertebrae and are placed immediately superficial to the rotatores. They are found along the whole length of the vertebral column from the sacrum to the cervical vertebrae. The second group is the *semispinalis*, which miss four to six vertebrae and lie superficial to the multifidus. These muscles are in the thoracic and cervical region. The highest is called the *semispinalis capitis* which is inserted into the occipital bone between the superior and inferior nuchal lines.

Because the rotatores, multifidus and semispinalis muscles run upwards and inwards from transverse processes to spinous processes they are collectively known as the *transversospinalis* group of muscles.

(3) *Superficially placed longitudinal muscles with a long interval between origin and insertion*

These muscles are known collectively as the *erector spinae (sacrospinalis)*. From the common origin on the back of the sacrum (superficial to multifidus) and adjacent iliac crest three columns arise—the medial or *spinalis* (from spines to spines), intermediate or *longissimus* (from common origin to transverse processes and ribs), and the lateral or *iliocostalis* (from common origin to ribs near their angles). Special slips of these muscles are continued up into the neck. Each of these columns is further subdivided according to their level in relation to the vertebral column, e.g. *iliocostalis lumborum, iliocostalis thoracis*, etc. *Longissimus capitis* is attached to the mastoid process deep to splenius capitis and sternocleidomastoid.

Nerve supply: All these muscles are supplied segmentally by the dorsal rami of the spinal nerves of their particular region.

Actions: If the muscles of both sides contract, they extend the head and the trunk. If the muscles of one side contract, they produce lateral bending (flexion) of the trunk. Lateral bending is always accompanied by some rotation to the opposite side. Muscles which pass upwards and outwards, e.g. splenius capitis, longissimus, rotate the head or vertebrae to the same side. Muscles passing upwards and inwards, e.g. transversospinales, rotate the vertebrae to the opposite side. Note that flexion of the trunk from the upright position is controlled by the extensor muscles, an example of relaxation of antagonists controlling a movement due to gravity.

Thoracolumbar (lumbar) fascia
Three lamellae, fused laterally along a line from 12th rib to iliac crest. *Posterior lamella* attached to tips of spinous processes of lumbar and sacral vertebrae, *intermediate lamella* to tips of lumbar transverse processes, *anterior lamella* to front of transverse process (Fig. 34). Posterior lamella continues upwards over thorax attached to thoracic spinous processes and angles of ribs, fades out over lower part of neck. Extensor muscles of vertebral column lie between posterior and middle lamellae, quadratus lumborum between middle and anterior lamellae.

The suboccipital muscles

These form a group of small muscles deep to the semispinalis capitis and connect the upper two cervical vertebrae to one another or to the head. All supplied by the dorsal ramus of the 1st cervical nerve (Fig. 26).

Inferior oblique
  Origin: Spine of axis.

  Insertion: Back of transverse process of atlas.

  Action: By rotating atlas, and therefore skull, turns head to same side.

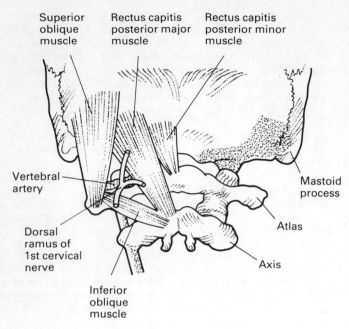

Superior
oblique
muscle

Rectus capitis
posterior major
muscle

Rectus capitis
posterior minor
muscle

Vertebral
artery

Mastoid
process

Dorsal
ramus of
1st cervical
nerve

Atlas

Axis

Inferior
oblique
muscle

Fig. 26 Suboccipital region.

Superior oblique
   Origin: Transverse process of atlas.

   Insertion: Lateral part of occipital bone between superior and inferior nuchal lines.

   Action: Extends and laterally bends head at atlanto-occipital joint.

Rectus capitis posterior major
   Origin: Spine of axis.

   Insertion: Lateral part of occipital bone inferior to inferior nuchal line.

Action: Extends head.

Rectus capitis posterior minor
   Origin: Tubercle of posterior arch of atlas.

   Insertion: Medial part of occipital bone inferior to inferior nuchal line (medial to and overlapped by rectus major).

   Nerve supply: Suboccipital nerve (dorsal ramus of C. 1).

   Action: Extensor of head.

   Note: The posterior recti, lateral recti and the anterior recti have a special function in fixing the skull firmly on the atlas during rotation.

Suboccipital triangle (Fig. 26)
The sides are rectus capitis posterior major, superior and inferior oblique muscles. Floor contains posterior arch of atlas and vertebral artery piercing posterior atlanto-occipital membrane. Greater occipital nerve (dorsal ramus of C. 2) appears between atlas and axis, pierces semispinalis capitis and runs upwards into scalp. Dorsal ramus of 1st cervical nerve is between vertebral artery and posterior arch of atlas and supplies all the muscles in triangle. It usually has no cutaneous branch.

THE MUSCLES OF THE UPPER LIMB

Included in this section are muscles which may appear to belong to the trunk, e.g. trapezius, levator scapulae and rhomboids. The clavicle and scapula (pectoral girdle), however, belong to the limb, so that most of the muscles attached to these bones act on the limb.

   The muscles of the upper limb form the following groups: (1) Muscles connecting limb to trunk: (a) posterior group, connecting limb to vertebral column, (b) anterior group, connecting limb to thoracic wall; (2) Muscles of the shoulder; (3) Muscles of the upper arm; (4) Muscles of the forearm; (5) Muscles of the hand.

(1) (a) POSTERIOR MUSCLES CONNECTING LIMB TO VERTEBRAL COLUMN

Trapezius
An extensive triangular sheet covering the back of the neck and the upper part of the back.

Origin: Medial one-third of superior nuchal line of occipital bone, external occipital protuberance, ligamentum nuchae (p. 6), spines and supraspinous ligaments of all the thoracic vertebrae.

Insertion: Lateral one-third of posterior border of clavicle (descending fibres), medial border of acromion (transverse fibres) and upper border of spine of scapula (ascending fibres).

Nerve supply: Motor—spinal part of accessory nerve; sensory (proprioceptive)—ventral rami of 3rd and 4th cervical nerves.

Actions: Elevates shoulder (upper fibres) (shrugging); braces shoulder back (middle fibres); with serratus anterior, lower fibres rotate scapula so that glenoid cavity faces upwards in full abduction of upper limb (p. 19); one trapezius rotates head to opposite side and both trapezii extend head; weight of upper limb transferred through clavicle, scapula and trapezius to vertebral column.
Note: Anterior border of trapezius forms posterior boundary of lateral region (posterior triangle) of neck.

Latissimus dorsi
A thin sheet covering the lower part of the back.

Origin: Spines and supraspinous ligaments of lower six thoracic vertebrae under cover of lower part of trapezius; by thoracolumbar fascia from spines of all lumbar vertebrae; outer lip of posterior part of crest of ilium; muscular slips join its deep surface from lower four ribs and from inferior angle of scapula. The iliac fibres form the posterior boundary of the lumbar triangle (of Petit).

Insertion: Passes upwards and laterally and narrows to a flat tendon which winds round lower border of teres major to be inserted into floor of intertubercular groove between teres major medially and pectoralis major laterally.

Note: The muscle twists on itself as it ascends to humerus and curves round the teres major.

Nerve supply: Thoracodorsal nerve from posterior cord of brachial plexus (C. 6, 7, 8).

Action: Extension and medial rotation of arm at shoulder joint from anatomical position; if upper limbs raised and fixed, latissimus dorsi elevates body, as in climbing a rope.

Levator scapulae

Origin: Posterior tubercles of transverse processes of upper four cervical vertebrae.

Insertion: Vertebral border of scapula from superior angle to root of spine.

Nerve supply: C. 3, 4, 5 (ventral rami).

Action: Assists upper fibres of trapezius in elevating scapula.

Rhomboids

These form a small thick sheet of muscle extending downwards from the vertebral spines to the vertebral border of the scapula. They act together in bracing back the shoulders. With levator scapulae they lie under cover of the trapezius.

Rhomboid major arises from the spines of the 2nd to the 5th thoracic vertebrae and is inserted into the vertebral border of the scapula from the root of the spine to the inferior angle.

Rhomboid minor extends from the lower part of the ligamentum nuchae and the spines of C. 7 and T. 1 to the vertebral border of the scapula opposite the root of the spine.

Nerve supply: The dorsal scapular nerve (C. 5), from brachial plexus in the neck, having passed under and supplied the levator scapulae, passes also deep to these two muscles and supplies them on their deep surface.

(1) (b) MUSCLES CONNECTING LIMB TO THORACIC WALL

Pectoralis major

Origin: (1) The medial half of the anterior surface of the clavicle (*clavicular head*). (2) The front of the sternum and upper six costal cartilages (*sternocostal head*). (3) Rectus sheath (i.e. external oblique aponeurosis).

Insertion: The whole length of the lateral lip of the intertubercular groove. Tendon of insertion is folded to become bilaminar. Clavicular head forms anterior lamina; sternocostal head twists under this to form posterior lamina; lowest fibres of origin inserted as high as capsule of shoulder joint.

Nerve supply: Lateral (C. 6, 7) and medial (C. 8, T. 1) pectoral nerves.

Action: Adducts and medially rotates arm at shoulder joint. Clavicular head flexes and sternocostal head extends arm from flexed position. With arm fixed in abduction may raise ribs (accessory muscle of inspiration).

Pectoralis minor
Lies under cover of pectoralis major.

Origin: 3rd, 4th and 5th ribs near junction with cartilages.

Insertion: Medial border of coracoid process.

Action: Pulls scapula forwards and depresses shoulder.

Nerve supply: Medial pectoral (C. 8, T. 1).

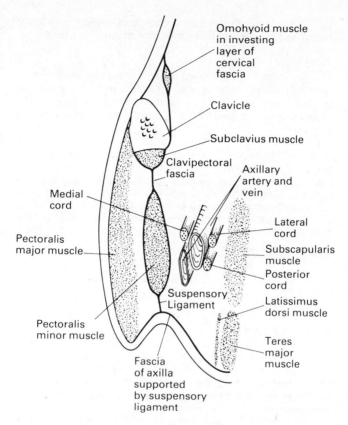

Fig. 27 Vertical section through left axilla.

Clavipectoral fascia (Fig. 27)
Fascia attached to clavicle around subclavius passing down
to upper border of pectoralis minor; extends from anterior
intercostal membrane to coracoid process. Pierced by acromiothoracic artery, lateral pectoral nerve, cephalic vein and
lymphatic vessels. Fascia encloses pectoralis minor and
becomes *suspensory ligament of axilla*—hence hollow in
armpit if clavicle be raised.

The fascia of the pectoral muscles contains important lymphatic plexuses which drain the mammary gland.

Subclavius

This arises by a tendon from the upper surface of the 1st rib near its cartilage, and spreads laterally and upwards to its insertion into the groove on the inferior surface of the clavicle. It is supplied by the nerve to the subclavius from C. 5 and 6.

Action: Depressor of clavicle, and stabilizes the clavicle for movements of arm.

Note: The axillary vessels and the brachial plexus lying on the 1st rib are separated from the clavicle by the subclavius muscle. The nerve to the subclavius is given off from the upper trunk of the plexus, runs down in front of the plexus and the subclavian vein, and often gives off the important accessory phrenic nerve (p. 272).

Serratus anterior

Origin: By digitations from outer surfaces of the upper eight ribs.

Insertion: Ventral aspect of whole length of vertebral border of scapula.

The first digitation arises from the 1st and 2nd ribs, and is inserted into the upper angle of the scapula, the next three arise from the 2nd, 3rd and 4th ribs, and spread out as thin sheets inserted along the vertebral border, and the lowest four from ribs 5 to 8 converge on to the inferior angle of the scapula.

Action: Draws scapula forwards as in lengthening the reach. Lower fibres rotate scapula so that glenoid fossa faces upwards, thus increasing abduction and flexion of the upper limb at the shoulder to 180°. May be an accessory muscle of respiration.

Nerve supply: The long thoracic nerve (C. 5, 6, 7), which runs vertically down the outer surface of the muscle (which is the medial wall of the axilla) deep to its fascia. Nerve lies just behind midaxillary line.

(2) MUSCLES OF THE SHOULDER

The *deltoid* is a large muscle covering the shoulder joint and giving to the shoulder its characteristic outline. Acromial fibres multipennate for greater power.

Origin: (1) The lateral one-third of the anterior border of the clavicle. (2) The outer border of the acromion. (3) The lower edge of the spine of the scapula.

Insertion: By a tendon into a well marked eminence on the outer side of the humerus nearly half-way down the bone.

Nerve supply: Axillary (circumflex) nerve (C. 5, 6), which winds round medial side of surgical neck of humerus and passes backwards inferior to the subscapularis, capsule of shoulder joint and teres minor. Downward dislocation of head of humerus or fracture of surgical neck can damage the nerve.

Action: Abducts arm at glenohumeral joint. Anterior fibres flex and medially rotate, and posterior fibres extend and laterally rotate arm at shoulder joint.

Between anterior border of deltoid and upper border of pectoralis major is the important *infraclavicular fossa* in which the cephalic vein lies before piercing the clavipectoral fascia. There are also lymph nodes in the fossa, and the fascia is pierced by the acromiothoracic artery and lateral pectoral nerve.

The short muscles surrounding the shoulder have individual actions, but act together to add stability to the joint and prevent the humerus moving in an unwanted direction.

Supraspinatus

Origin: Supraspinous fossa.

Insertion: Upper facet on greater tuberosity of humerus; tendon blends with upper part of capsule of shoulder joint.

Nerve supply: Suprascapular (C. 5, 6) from upper trunk of brachial plexus.

Action: Involved in initiation of abduction holding the head of the humerus down so that upward pull of deltoid is converted into a lateral pull.

Infraspinatus

Origin: Infraspinous fossa.

Insertion: Middle facet on greater tuberosity of humerus; capsule of shoulder joint.

Nerve supply: Suprascapular nerve.

Action: Lateral rotator

Teres minor

Origin: Upper part of dorsal aspect of lateral border of scapula.

Insertion: Lowest facet on greater tuberosity of humerus and adjoining part of shaft.

Nerve supply: Axillary nerve (C. 5, 6).

Action: Lateral rotator.

Teres major

Origin: Back of inferior angle and lateral border of scapula below teres minor.

Insertion: Front of humerus—medial lip of intertubercular groove. Lies edge to edge with subscapularis.

Nerve supply: Lower subscapular (C. 6, 7).

Action: Medial rotator and extensor of arm at shoulder joint. Adducts abducted arm against resistance.

Subscapularis
This muscle covers the front of the shoulder joint, and forms the greater part of the posterior wall of the axilla, which is completed below by teres major and latissimus dorsi.

Origin: The whole of the subscapular fossa, except the neck of the scapula, where the muscle is separated from bone by a bursa which communicates with shoulder joint. Multipennate fibres arise from fibrous septa in the muscle, which are attached to ridges in the subscapular fossa.

Insertion: Lesser tuberosity of humerus and shaft below this.

Nerve supply: Upper and lower subscapular nerves from posterior cord of plexus (C. 5, 6, 7).

Action: Medial rotator, and also fixator as described above.

Muscular walls of axillary space
Anterior: Pectoralis major superficially and clavicle, subclavius, clavipectoral fascia and pectoralis minor deep.

Medial: Serratus anterior and upper ribs.

Posterior: Subscapularis, teres major, latissimus dorsi.

Lateral: Humerus and coracobrachialis.

The *quadrilateral space* is an exit from the axilla through the posterior wall and is bounded above by the subscapularis, capsule of shoulder joint and teres minor, below by teres major, laterally by the humerus and medially by the long head of the triceps brachii. Axillary nerve and posterior circumflex humeral vessels traverse quadrilateral space.

(3) MUSCLES OF THE UPPER ARM

Coracobrachialis

Origin: Tip of the coracoid process in common with the tendon of short head of biceps brachii.

Insertion: Medial aspect of shaft of humerus about half-way down.

Nerve supply: Pierced by musculocutaneous nerve, which supplies it (C. 7).

Action: Flexor of humerus at shoulder joint.
Note: The coracobrachialis forms a ridge in the arm; the brachial artery lies behind it, and passes forwards across its insertion.

Biceps brachii

This muscle extends from the scapula to the radius, thus crossing and acting on the shoulder and elbow joints.

Origin: By two heads, *short head* from the tip of the coracoid process with coracobrachialis and *long head* from the supraglenoid tubercle of the scapula within the fibrous capsule of the shoulder joint (p. 18).
The two heads joint together about the middle of the arm.

Insertion: (1) Principally by a strong tendon into the posterior half of the tuberosity of the radius. Tendon rotates so that anterior surface at elbow faces laterally at its insertion. (2) By a fascial band (*bicipital aponeurosis*) from the medial side of the tendon into the deep fascia of medial side of forearm, and so to subcutaneous border of ulna.

Nerve supply: Musculocutaneous (C. 5, 6) as it passes deep to the muscle.

Action: Supinator and flexor of forearm; may assist in strong flexion of humerus at shoulder joint; long head enhances stability of shoulder joint.
Note: (1) Brachial artery and median nerve lie in a visible groove along medial border of the muscle.(2) Basilic vein and medial

cutaneous nerve of forearm are superficial to bicipital aponeurosis, and brachial artery and median nerve lie under cover of it. In each case nerve is medial to vessel.

Brachialis
Lies deeply under cover of lower half of biceps.

Origin: Lower half of front of shaft of humerus, and medial intermuscular septum. Fibres converge downwards to a strong tendon.

Insertion: Coronoid process and tuberosity of ulna.

Nerve supply: Musculocutaneous (C. 5, 6).

Action: Flexor of forearm at elbow.

The *triceps brachii* constitutes the whole of the muscular mass of the back of the arm, arises from the back of the humerus, and by an additional head from the scapula, and is inserted by a conjoined tendon into the olecranon of the ulna.

Origin: Long head from the infraglenoid tubercle of the scapula outside capsule of shoulder joint; *lateral head* from a ridge on back of humerus above groove for radial nerve, extending almost as high as teres minor; *medial head* from the whole width of the posterior surface of the humerus below the groove extending as high as the teres major, and from medial and lateral intermuscular septa.
Long and lateral heads join to form a massive tendon; this lies superficial to medial head, and the fibres of medial head are inserted into its deep aspect.

Insertion: The tendon is attached to the posterior part of the superior surface of the olecranon.

Nerve supply: Radial (C. 7, 8).
Note: Medial head receives two branches from radial nerve; one branch, sometimes called *ulnar collateral nerve*, is given off in the axilla and enters it low down; the second branch is given off in the groove for radial nerve and runs down in the medial head to

supply it and the anconeus. The long head is supplied in the axilla, and the lateral head in the groove.

Action: Extension of forearm at elbow. Long head supports shoulder joint in abduction.

(4) MUSCLES OF THE FOREARM

Deep fascia of the forearm
This encircles the muscles of the forearm, sends septa in between them, and is attached posteriorly to the subcutaneous border of the ulna.

The *flexor retinaculum* (a thickened part of the deep fascia) stretches transversely across the front of the wrist from the tubercles of scaphoid and trapezium laterally to the pisiform and hook of hamate medially.

The *extensor retinaculum* (another thickening of deep fascia) stretches very obliquely downwards and medially from the distal part of the radius to the pisiform and triquetral. It is therefore unaffected by pronation and supination.

Superficial muscles of the front of the forearm

These arise by a common flexor origin from the front of medial epicondyle of the humerus and from the fascia over them and the septa between them; some of them have additional origins.

Pronator teres
Origin: Common flexor origin and lower part of supracondylar ridge; an additional deep origin from medial aspect of coronoid process of ulna.

Insertion: Middle of lateral aspect of radius, i.e. region of greatest convexity.

Nerve supply: Median (C. 6).

Action: Pronation of forearm.
Note: Nerve to pronator teres is given off in cubital fossa (p. 85).

Median nerve passes between two heads and ulnar artery deep to both heads of muscle.

Flexor carpi radialis
  Origin: Common flexor origin.

  Insertion: Front of base of 2nd metacarpal with slip to 3rd.
  Note: Tendon runs in special synovial sheath deep to flexor retinaculum and in groove on trapezium.

  Nerve supply: Median (C. 6, 7).

  Action: Abduction (radial deviation) and flexion of hand at wrist; flexor only with flexor carpi ulnaris; abductor only with radial extensors. Fixator of wrist in certain movements of fingers.

Palmaris longus (may be absent)
  Origin: Common flexor origin.

  Insertion: Into proximal end of palmar aponeurosis (p. 91); tendon is superficial and adherent to flexor retinaculum.

  Nerve supply: Median (C. 6, 7).

  Action: Tensor of palmar fascia and flexor of hand at wrist.

Flexor carpi ulnaris
  Origin: Common flexor origin; medial margin of olecranon of ulna; aponeurosis attached to upper two-thirds of subcutaneous border of ulna.

  Insertion: Pisiform bone; by extensions to hook of hamate (*pisohamate ligament*), base of 5th metacarpal (*pisometacarpal ligament*) and front of flexor retinaculum.
  Note: The true insertion is into the metacarpal bone (compare flexor carpi radialis); the others may be regarded as interruptions of the tendon.
  Nerve supply: Ulnar (C. 7, 8).

Action: Adduction (ulnar deviation) and flexion of hand at wrist; flexion only with flexor carpi radialis (and palmaris longus); adduction only with extensor carpi ulnaris.

Flexor digitorum superficialis (sublimis)
Lies in a plane intermediate between the above muscles and the deep muscles.

Origin: Common flexor origin; medial ligament of elbow and highest point of medial margin of coronoid process; oblique line on anterior surface of shaft of radius and fibrous arch between bones.

Insertion: By four tendons each of which after splitting to allow tendon of flexor digitorum profundus to pass

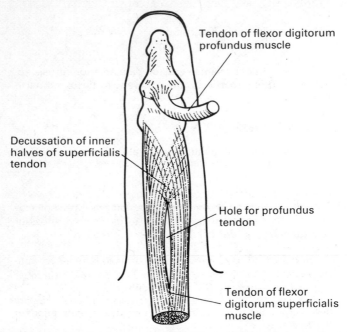

Tendon of flexor digitorum profundus muscle

Decussation of inner halves of superficialis tendon

Hole for profundus tendon

Tendon of flexor digitorum superficialis muscle

Fig. 28 Chiasma in tendon of superficial (sublimis) flexor tendon (profundus tendon is displaced).

through, joins together again, and splits again to go to the sides of the middle phalanx of medial four digits (Fig. 28).

Nerve supply: Median (C. 7, 8).

Action: Flexion of middle phalanx at proximal interphalangeal joint and proximal phalanx at metacarpophalangeal joint; weak flexor at wrist and elbow joints.

Deep muscles of the front of the forearm

Flexor digitorum profundus
Large muscle which gives bulk to medial part of forearm.

Origin: Front and medial surface of shaft of ulna above pronator quadratus and adjacent interosseous membrane.

Insertion: By four tendons each of which is attached to front of base of terminal phalanx of medial digits; these tendons perforate those of the superficial flexor (Fig. 28).

Nerve supply: Anterior interosseous (median) to index and middle fingers, ulnar to ring and little fingers (C. 7, 8) similar to nerve supply of lumbricals.

Action: Flexion of terminal phalanx at distal interphalangeal joint and secondarily of other joints of finger and wrist.

Tendons and sheaths of long flexors of fingers
(1) Common synovial sheath under flexor retinaculum for flexors superficialis and profundus (Fig. 29). Flexor superficialis tendons in pairs, middle and ring fingers anterior to index and little finger tendons. Profundus tendons side by side but only index tendon separated at this level (Fig. 29).
(2) This sheath extends about 2.5 cm proximal and distal to the flexor retinaculum and along little finger tendon to base of terminal phalanx of that digit, but not along other digits.
(3) Synovial sheaths along whole length of index, middle and ring fingers, not continued into palm.

(4) Fibro-osseous tunnels formed along all the medial digits each containing the two tendons and the synovial sheath.

(5) Vincula are small synovial folds (like mesenteries) connecting tendons to periosteum of metacarpal or phalanx or to ligaments of interphalangeal joints; they carry supplementary blood supply from digital arteries to the tendons.

(6) On the front of the proximal phalanx the tendon of flexor superficialis splits and joins again to form a spiral ring around the tendon of the profundus. It then splits again, as described, to be attached to the sides of the middle phalanx.

Flexor pollicis longus
Origin: Front of shaft of radius—above pronator quadratus and below anterior oblique line.

Insertion: Front of base of distal phalanx of thumb.

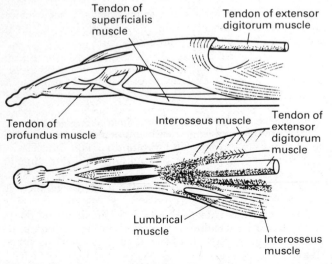

Fig. 29 Dorsal extensor expansion.

Nerve supply: Anterior interosseous of median (C. 7, 8).

Action: Flexion at interphalangeal joint of thumb and secondarily at other joints which it crosses (metacarpophalangeal and carpometacarpal).

Note: Tendon runs in separate synovial sheath deep to flexor retinaculum, and this sheath is prolonged to base of distal phalanx. Communicates at wrist level in 50% of cases with common synovial sheath of long digital flexors.

Pronator quadratus
A quadrilateral transverse muscle at lower end of forearm.

Origin: Ridge on front of lower quarter of shaft of ulna.

Insertion: Distal quarter of front and medial side of shaft of radius.

Nerve supply: Anterior interosseous of median (C. 7, 8).

Action: By pulling lower end of radius round ulna it pronates the forearm and hand.

Note: The cubital fossa is bounded by pronator teres medially and brachioradialis laterally; its floor is formed by brachialis and supinator. In the subcutaneous tissue lying over the fossa there are important veins used for obtaining blood and giving injections. The tendon of biceps brachii, brachial artery and median nerve lie in the floor of the fossa from lateral to medial side (the artery is used for obtaining the blood pressure). The radial nerve is lateral deep to brachioradialis on brachialis.

Superficial muscles of the back of the forearm

These are extensors and supinators, except for the brachioradialis. The common extensor origin is on the anterior part of the lateral epicondyle of the humerus; thus if forearm is supinated and extended, muscles arising from it curve backwards into forearm and produce characteristic posterior hollow medial to the muscles in which lies head of radius. In working position of almost full pronation, extensor muscles pass straight into forearm.

Brachioradialis
  Origin: Upper two-thirds of lateral supracondylar ridge of humerus.

  Insertion: Into outer side of styloid process of radius.

  Nerve supply: Radial (C. 5, 6) in the arm.

  Action: Flexion of elbow in position midway between supination and pronation.

Extensor carpi radialis longus
  Origin: Lower one-third of lateral supracondylar ridge.

  Insertion: Back of base of 2nd metacarpal.

  Nerve supply: Radial (C. 6, 7) in arm.

  Action: Extension and abduction of hand at wrist; fixation of wrist when long flexors are acting on fingers (synergic action). The next four muscles arise from the common origin.

Extensor carpi radialis brevis
  Origin: Common extensor origin.
  Insertion: Back of base of 3rd metacarpal.

  Nerve supply: Posterior interosseous (C. 6, 7) in arm.

  Action: Extension and abduction of hand at wrist. In extension both radial carpal extensors act with extensor carpi ulnaris. In abduction they act with flexor carpi radialis and abductor pollicis longus.

Extensor digitorum
  Origin: Common extensor origin.

  Insertion: By four flat tendons which spread out on the back of the hand and go to medial four digits. Joined in variable manner to each other by oblique intertendinous connexions. The flat tendon on the back of each finger is

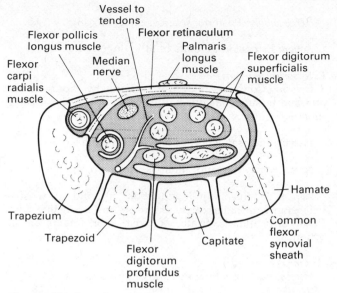

Fig. 30 Tendons deep to flexor retinaculum to show synovial sheaths.

the *dorsal* or *extensor expansion*; it receives on the lateral side a tendon of a lumbrical (p. 93) and on each side an interosseus (p. 93). Then its central part goes to the back of the base of the middle phalanx, and its margins as two slips to the back of the base of the terminal phalanx (Fig. 30).

Note: The tendons on the back of the hand are incorporated in the deep fascia and form with it a complete covering of the back of the hand.

Nerve supply: Posterior interosseous (C. 7, 8).

Action: Extension at metacarpophalangeal and interphalangeal joints of digits and secondarily of hand at wrist.

Extensor digiti minimi
  Origin: Common extensor origin.

  Insertion: Into medial side of extensor expansion of little finger.

Nerve supply: Posterior interosseous (C. 7, 8).

Action: Extension at joints of little finger.

Extensor carpi ulnaris
   Origin: Common extensor origin; by aponeurosis attached to subcutaneous border of ulna with flexor carpi ulnaris.

   Insertion: Dorsal aspect of base of 5th metacarpal.

   Nerve supply: Posterior interosseous (C. 6, 7).

   Action: Extension and adduction of hand at wrist—extension with radial carpal extensors and adduction with flexor carpi ulnaris; fixator of wrist when fingers are in action.

Anconeus
   Origin: Back of lateral epicondyle of humerus.

   Insertion: Triangular area on lateral aspect of proximal end of ulna.

   Nerve supply: Nerve to medial head of triceps brachii which runs in substance of triceps to anconeus.

   Action: Extends forearm at elbow and may pull ulna laterally and posteriorly in pronation (pronation and supination, see p. 22).
   Note: The anconeus forms a thin triangular sheet covering the head of the radius.

Deep muscles of the back of the forearm

These are supinator and long extensors of the thumb and index finger.

Supinator
   Origin: Deeper (transverse) fibres from ulna–supinator crest and area just below radial notch; more superficial

(oblique) fibres from back of lateral epicondyle, lateral ligament of elbow and annular ligament.

Insertion: Lateral surface of upper part of shaft of the radius extending on to its anterior and posterior surfaces.

Nerve supply: Posterior interosseous nerve (C. 6) which passes backwards between the two heads.

Action: Supinates the pronated radius.

Abductor pollicis longus
   Origin: Proximal part of back of ulna; back of middle of radius; interosseous membrane.

Insertion: Lateral aspect of base of metacarpal of thumb.

Nerve supply: Posterior interosseous (C. 7, 8).

Action: Abduction and extension of metacarpal of thumb at carpometacarpal joint; abductor of hand at wrist.

Extensor pollicis brevis
   Origin: Back of radius distal to abductor pollicis longus; adjacent part of interosseous membrane.

Insertion: Back of base of proximal phalanx of thumb.

Nerve supply: Posterior interosseous (C. 7, 8).

Action: Extension of proximal phalanx of thumb and secondarily abduction of hand at wrist.

Extensor pollicis longus
   Origin: Back of ulna distal to abductor pollicis longus and adjacent part of interosseous membrane. Long tendon changes direction around dorsal (Lister's) tubercle of radius. More distally forms with previous two tendons the *snuff-box* in which are palpable the base of thumb metacarpal, trapezium, scaphoid, radial styloid and radial artery

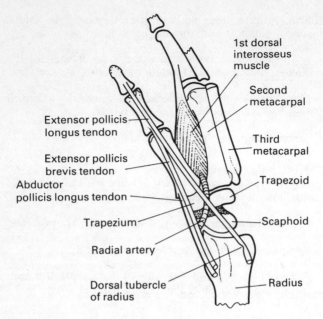

Fig. 31 Contents of right 'snuff-box'.

(Fig. 31). Superficial part of radial nerve and beginning of cephalic vein are subcutaneous.

Insertion: Back of base of terminal phalanx of thumb.

Nerve supply: Posterior interosseous (C. 7, 8).

Action: Extension of thumb, and secondarily of hand at wrist.

Extensor indicis
Origin: Back of ulna distal to extensor pollicis longus and interosseous membrane.

Insertion: Joins medial side of dorsal expansion of index finger.

Nerve supply: Posterior interosseous (C. 7, 8).

Note: Compartments under extensor retinaculum:(1) Lateral aspect of radius—abductor pollicis longus and extensor pollicis brevis. (2) Dorsal aspect of radius (from lateral to medial side): (a) extensors carpi radialis longus et brevis; (b) extensor pollicis longus; (c) extensor digitorum and extensor indicis. (3) Between radius and ulna—extensor digiti minimi. (4) Dorsal aspect of ulna—extensor carpi ulnaris. Each of the six compartments has its own synovial sheath extending proximally (about 7 mm) and distally (about 15 mm) beyond the extensor retinaculum.

(5) MUSCLES OF THE HAND

The *palmar aponeurosis* is the deep fascia of the palm of the hand. Over the thenar and hypothenar eminences it is thin, but in the hollow of the palm it is thick. This central part is a continuation of palmaris longus. It is attached proximally to the flexor retinaculum; distally it splits into slips to the four digits which blend with the fibrous flexor sheaths of the fingers. From its deep surface laterally a septum passes into the palm to the front of the middle metacarpal, separating off a lateral *thenar space* containing the adductor pollicis muscle from a more medial *midpalmar space* containing the long flexor tendons. The long tendons of the index finger commonly lie in the thenar space.

The muscles of the hand are in three groups: (1) muscles of the thumb forming *thenar eminence*, (2) muscles of the little finger forming *hypothenar eminence*, (3) palmar muscles—adductor pollicis, interossei and lumbricals.

The thumb and little finger each have three superficial muscles, while the thumb has an extra deep muscle, adductor pollicis. The superficial groups of the thumb and little finger closely resemble one another. In addition to their bony origins they are attached to the flexor retinaculum.

Origin: Abductor pollicis brevis (lateral), *flexor pollicis brevis* (medial) and *opponens pollicis* (deep to the other two) from scaphoid and trapezium; corresponding muscles of hypothenar eminence (*abductor digiti minimi*-medial, *flexor digiti minimi*-lateral and *opponens digiti minimi*) from pisiform and hook of hamate.

Insertion: Abductor and flexor pollicis brevis to the lateral side of base of proximal phalanx of thumb; abductor and flexor digiti minimi to medial side of the base of proximal phalanx of little finger; opponens pollicis to lateral side of shaft of 1st metacarpal and opponens digiti minimi to medial side of shaft of 5th metacarpal.

Nerve supply: Thenar muscles by median nerve (T. 1) and hypothenar muscles by ulnar nerve (T. 1).

Actions: Flexor pollicis brevis—flexion of thumb at meta-carpophalangeal and carpometacarpal joints; abductor pollicis brevis—abduction of thumb at same joints; opponens pollicis—opposition of thumb (p. 25). Flexor digiti minimi—flexion at metacarpophalangeal joint of little finger; abductor digiti minimi—abduction at same joint; opponens digiti minimi—longitudinal rotation of 5th metacarpal towards middle of palm.

Note: The so-called *deep head* of the *flexor pollicis brevis* is really the 1st palmar interosseus.

Palmaris brevis consists of a few transverse fibres in the superficial fascia of the hypothenar eminence; supplied by the ulnar nerve.

Adductor pollicis
Origin: Oblique head: front of bases of 2nd and 3rd metacarpals, and adjacent part of carpus (capitate and trapezoid); *transverse head:* distal two-thirds of front of 3rd metacarpal.

Insertion: Two heads join and are inserted into medial aspect of base of the proximal phalanx of the thumb; some fibres go to lateral aspect of base.

Nerve supply: Ulnar nerve.

Action: Adduction of thumb, i.e. movement toward index finger at right angles to plane of palm.

Note: There are sesamoid bones in common tendons of muscles inserted into medial and lateral sides of proximal phalanx of thumb.

Lumbricals
Four slender muscles associated with the tendons of the flexor digitorum profundus.

Origin: 1st and 2nd, from the radial sides of 1st and 2nd tendons; 3rd and 4th, from the adjacent sides of 2nd and 3rd and 3rd and 4th tendons.

Insertion: Each lumbrical tendon passes distally on the palmar aspect of the deep transverse metacarpal ligament and round the radial side of the appropriate digit to be inserted into the dorsal expansion (p. 87) on the middle phalanx (Fig. 29).

Nerve supply: 1st and 2nd by digital branches of median; 3rd and 4th by deep branch of ulnar.

Action: Flexion at metacarpophalangeal joint and extension at interphalangeal joints.

Interossei
These are four dorsal and four palmar, but a part of each dorsal interosseus is in series with the palmar interossei and can be seen in a dissection of the palm. All eight are supplied by the ulnar nerve.

Dorsal interossei
   Origin: From the adjacent sides of two metacarpals.

Insertion: 1st, into radial side of base of proximal phalanx of index finger and into dorsal expansion.
   2nd, into radial side of base of proximal phalanx of middle finger and dorsal expansion.
   3rd, into ulnar side of base of proximal phalanx of middle finger and dorsal expansion.
   4th, into ulnar side of base of proximal phalanx of ring finger and dorsal expansion.

Action: (1) Abduction of extended digits from an axis represented by middle finger. (2) Flexion at metacarpophalangeal joint and extension at interphalangeal joints (cf. lumbricals).

Palmar interossei
  Origin: 1st, from ulnar side of base of 1st metacarpal.
2nd, from ulnar side of shaft of 2nd metacarpal.
3rd, from radial side of shaft of 4th metacarpal.
4th, from radial side of shaft of 5th metacarpal.

  Insertion: 1st and 2nd, into ulnar side of proximal phalanx
of thumb and index finger respectively.
  3rd and 4th, into radial side of proximal phalanx of ring
and little fingers.

  Action: (1) Adduction of digits to axis represented by
middle finger. (2) Flexion at metacarpophalangeal joints
and extension at interphalangeal joints (like dorsal interossei and lumbricals).
  Note: Layers of palm: (1) palmar fascia, (2) superficial palmar
arterial arch and median nerve, (3) superficial and deep flexor
tendons and lumbricals, (4) adductor pollicis, deep palmar arterial
arch and ulnar nerve, (5) interossei.

THE MUSCLES OF THE LOWER LIMB

MUSCLES OF THE ILIAC REGION

The *psoas major* is a large fusiform muscle extending from
the lumbar region of the vertebral column along the pelvic
brim and passing deep to the inguinal ligament to the
thigh.

  Origin: From adjacent borders of vertebrae T. 12 to L. 5
and intervening intervertebral discs; from fibrous arches
spanning lumbar vessels and sympathetic rami, as these lie
in the hollow of the vertebral bodies; the medial half of the
front of all the lumbar transverse processes.

  Insertion: By a tendon into lesser trochanter of femur.

  Nerve supply: L. 2, 3.

  Action: Flexion of thigh at hip; flexion of trunk at hip;
doubtful medial rotator of extended thigh.
  Note: (1) Fascial psoas sheath encloses muscle and extends to
insertion. (2) Lumbar nerves form lumbar plexus in substance of

muscle. (3) Forms important part of posterior abdominal wall with such structures as ureter, gonadal and colic vessels and inferior vena cava lying on it. (4) In thigh the tendon lies on front of hip joint (with a bursa intervening), and lies behind the femoral artery.

Psoas minor (when present) arises from lower border of body of T. 12, and goes to iliopubic eminence.

Iliacus
   Origin: Greater part of iliac fossa, extending on to sacrum. A large bursa lies deep to it in lower part of iliac fossa.

   Insertion: Lateral aspect of tendon of psoas and femur inferior to greater trochanter.

   Nerve supply: Femoral nerve (L. 2, 3).

   Action: With psoas flexes thigh at hip.
The iliacus has a fascial sheath similar to that of psoas.

MUSCLES OF THE THIGH

The deep fascia (fascia lata) of the thigh
This surrounds the muscles and sends septa in between the principal groups; it is very strong, especially on the lateral aspect of the thigh. It is attached above to the iliac crest, inguinal ligament, ischiopubic ramus and sacrum. It is pierced obliquely by the *saphenous opening* 3 cm below and lateral to the pubic tubercle; from this opening a deeper layer passes deeply to become continuous with the fascia over the pectineus, and the medial intermuscular septum; between these two layers is the femoral sheath. Saphenous opening covered by *cribriform fascia*, and transmits great saphenous vein, lymphatics and small arteries.
   Inferiorly, the fascia is attached to the tibial condyles and to the sides of the patella, blending with the patellar retinacula (p. 34). On the lateral side, the fascia lata is thickened to form the *iliotibial tract* which receives the insertions of three-quarters of the gluteus maximus and the whole of the tensor fasciae latae. The tract is inserted into front of lateral condyle of tibia.

The anterior thigh muscles

These are the large *quadriceps femoris* with the small articularis genus and the sartorius. All are supplied by the femoral nerve. The quadriceps is the extensor of the leg at the knee and consists of three muscles arising from the front and sides of the femur, and one from the hip bone, all inserted through the patella into the front of the tibia.

Vastus lateralis
   Origin: By strong aponeurosis from upper half of lateral lip of linea aspera, extending up to side and front of base of greater trochanter.

   Insertion: Muscular fibres extend downwards and medially to join common tendon of quadriceps femoris; lower tendinous fibres (retinacula) are inserted into side of patella and lateral part of ligamentum patellae.

Vastus intermedius
This lies deep to rest of muscles.

   Origin: Upper two-thirds of front and lateral aspect of shaft of femur.

   Insertion: Deep aspect of common tendon of quadriceps femoris.

Vastus medialis
   Origin: Medial lip of linea aspera; spiral line inferior to lesser trochanter from linea aspera to intertrochanteric line; medial supracondylar ridge; by muscular fibres from tendon of ischial part of adductor magnus (p. 99).

   Insertion: Into common tendon of quadriceps femoris; lower muscular fibres extend further distally than those of vastus lateralis and are horizontal to become attached to medial side of patella, preventing its lateral dislocation (Fig. 32).
   Note: The vastus medialis forms the lateral wall of the adductor (subsartorial) canal.

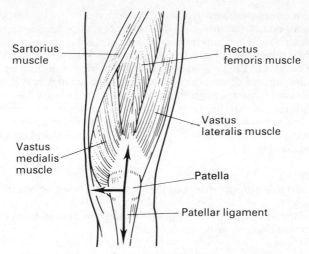

Fig. 32 Insertion of left vastus medialis into medial border of patella, preventing dislocation laterally when quadriceps contracts.

Rectus femoris
Fusiform bipennate muscle on a plane superficial to the vasti.

Origin: By two heads: straight head from anterior inferior iliac spine; reflected head from upper margin of acetabulum blending with capsule of hip joint.

Insertion: By tendon into upper border of patella; this tendon receives aponeurotic fibres of vasti, the whole forming the common *tendon of the quadriceps.*

The ultimate insertion of these four muscles is into tubercle of tibia. The patella is a sesamoid bone in the tendon of the quadriceps femoris; the tendon is continued distal to the patella as the *ligamentum patellae* and at the sides of the patella as the *retinacula* of the knee joint.

The *articularis genus* consists of a few muscle fibres from the front of the femur to the suprapatellar bursa of the knee (p. 35). It pulls up the synovial pouch during extension.

Action of quadriceps femoris: All components extend leg at knee; rectus femoris also flexes thigh at hip. Quadriceps femoris is an antigravity muscle and prevents flexion at the knee when the weight of the body falls behind the transverse axis of the knee joint as in sitting down, standing up, going up and down stairs. A weak quadriceps may result in the knee giving way in these situations. Pull of quadriceps femoris on patella is in line of femur (oblique) while ligamentum patellae is vertical; horizontal fibres of vastus medialis indispensable in preventing lateral dislocation of patella (Fig. 32).

Sartorius
   Origin: Anterior superior iliac spine and bone below it.

   Insertion: Upper part of medial surface of tibia in front of semitendinosus and gracilis.

   Nerve supply: Superficial division of femoral (L. 2, 3).

   Action: Flexion, abduction and lateral rotation of thigh at hip and flexion of leg at knee.
   Note: The sartorius is narrow and ribbon-like, crossing the thigh obliquely, superficial to the quadriceps. It forms the lateral boundary of the *femoral triangle* which is above and medial to sartorius and is bounded above by inguinal ligament and medially by adductor longus. Muscles in floor of triangle are iliacus, psoas major and pectineus and triangle contains femoral sheath surrounding femoral vessels, and femoral nerve lateral to and outside the sheath. The sartorius forms the anterior wall of the *adductor (subsartorial) canal* containing continuation of femoral vessels and nerve and nerve to vastus medialis, forming lateral wall of canal.

The medial thigh muscles

These occupy the adductor compartment. They are supplied by the obturator nerve which may supply the pectineus.

Obturator externus
   Origin: Outer surface of obturator membrane, and adjacent rami of pubis and ischium.

Insertion: Trochanteric fossa on medial side of upper part of greater trochanter.

Nerve supply: Posterior division of obturator nerve which leaves pelvis by piercing the muscle (p. 282).

Action: Lateral rotation of thigh at hips.
Note: Tendon of obturator externus winds backwards inferior to and then behind neck of femur in contact with the bone.

Adductor magnus
Origin: Outer surface of ischiopubic ramus and of tuberosity of ischium.

Insertion: Pubic fibres run nearly horizontally laterally to upward extension of linea aspera (gluteal ridge); fibres from ramus of ischium run obliquely to linea aspera and upper part of medial supracondylar ridge; fibres from tuberosity form long vertical muscle with tendinous insertion into adductor tubercle above medial epicondyle of femur.

Nerve supply: Posterior division of obturator (L. 2, 3), but ischial head is supplied by sciatic (L. 4, 5).

Action: Adduction and extension at hip joint.
Note: Ischial head is said to be a hamstring muscle, hence its nerve supply. Femoral artery passes backwards through opening in adductor magnus between the adductor and ischial parts, medial to femoral shaft. Perforating arteries (p. 179) pass backwards through adductor magnus.

Adductor brevis
Lies immediately in front of magnus.

Origin: Outer surface of inferior pubic ramus above adductor magnus.

Insertion: Upper part of linea aspera and extension of this to lesser trochanter.

Nerve supply: Anterior division of obturator nerve.

Action: Adduction at hip joint.
Note: This muscle separates the two divisions of obturator nerve.

Adductor longus
Longer, thinner and more oblique than brevis and anterior to it.

   Origin: Front of body of pubis by round tendon.

   Insertion: Whole length of linea aspera.

   Nerve supply: Anterior division of obturator.

   Action: Adduction and flexion of thigh.
   Note: This muscle separates femoral and profunda femoris vessels. Medial border forms one side of femoral triangle.

Pectineus
Above and in same plane as adductor longus.

   Origin: Pectineal line of pubis and triangular area in front of this.

   Insertion: Line from linea aspera to lesser trochanter medial to insertion of adductor brevis.

   Action: Adductor and flexor of thigh.

   Nerve supply: Usually anterior division of femoral (L. 2, 3); may be accessory obturator (L. 3).
   Note: Pectineus and adductor longus form floor of femoral triangle. Adductor magnus and longus form floor of subsartorial canal.

Gracilis
A thin ribbon-like muscle running vertically down medial side of thigh.

   Origin: Inferior margin of body and inferior ramus of pubis.

   Insertion: Upper part of medial surface of shaft of tibia, behind sartorius and above semitendinosus.

Nerve supply: Anterior division of obturator.

Action: Adduction of thigh and flexion of leg at knee.

The muscles of the gluteal region

These are the three glutei, the tensor fasciae latae and the lateral rotators, a series of short muscles closely related to the hip joint.

The gluteus maximus forms the mass of the buttock.

Gluteus maximus

Origin: Small area on ilium above and behind posterior gluteal line; back of sacrum below and laterally; back of sacrotuberous ligament.

Insertion: Mostly into deep fascia and iliotibial tract. Lower deep fibres (about one-quarter of bulk of muscle) into gluteal ridge of femur.

Nerve supply: Inferior gluteal (L. 5, S. 1, 2).

Action: Extension and lateral rotation of thigh; extends trunk on thigh, important action in straightening up from bent position; important muscle in walking up stairs and up a slope (extension of flexed thigh); through iliotibial tract extends at knee if foot on ground, flexes at knee if foot off ground.

Note: (1) Very coarse loose fasciculi of muscle. (2) It covers posterior part of gluteus medius, and stout gluteal fascia covering this muscle. (3) In the upright position its lower border overlaps ischial tuberosity and sacrotuberous ligament, a bursa intervening. (4) It covers short posterior muscles of hip joint and sciatic nerve lying on them, upper parts of hamstrings and both sciatic foramina with the structures passing through them. (5) Its lower border, which is oblique, does not correspond to the *gluteal fold* which is transverse and an extension crease of hip joint like flexion creases of fingers.

Gluteus medius

Origin: From outer surface of ilium between posterior and anterior gluteal lines.

Insertion: Oblique line running diagonally forwards and downwards across outer aspect of greater trochanter, and an area below this.

Nerve supply: Superior gluteal (L. 4, 5, S. 1).

Action: Abduction of thigh or tilting upwards of opposite side of pelvis with limb on ground; in walking gluteus medius of limb on ground supports the pelvis and prevents it from dropping on the unsupported side.

Gluteus minimus
Origin: The outer surface of the ilium between the anterior and inferior gluteal lines.

Insertion: The front of the greater trochanter of the femur.

Nerve supply: Superior gluteal (L. 4, 5, S. 1).

Action: Abductor and medial rotator of thigh, and in walking assists gluteus medius.

Tensor fasciae latae
This is described here since it is derived from the gluteal muscles, it is supplied by the superior gluteal nerve and its action is to assist the gluteus maximus in its action on the iliotibial tract.

Origin: Outer surface of ilium below anterior part of iliac crest between tubercle of crest and anterior superior iliac spine.

Insertion: Iliotibial tract.

Nerve supply: Superior gluteal (L. 4, 5, S. 1).

Action: In addition to action on iliotibial tract, flexes and medially rotates thigh at hip joint.

Piriformis
Origin: From the front of the middle three pieces of the sacrum within the pelvis; leaves pelvis through greater sciatic foramen.

Insertion: Pointed posterior extremity of upper border of greater trochanter.

Nerve supply: Directly from 2nd and 3rd sacral nerves.

Obturator internus
Origin: From the obturator membrane inside the pelvis, the bone around the obturator foramen and a large area behind and above the foramen. Tendon leaves pelvis through lesser sciatic foramen, then bends at a right angle round edge of notch where there is a bursa.

Insertion: Medial surface of greater trochanter above trochanteric fossa.

Nerve supply: Nerve to obturator internus (L. 5, S. 1, 2).

Gemelli
These are two slips of muscle arising from the upper and lower margins respectively of the lesser sciatic notch. They are inserted into the sides of the tendon of the obturator internus of which they are detached portions and whose action they reinforce.
Inferior supplied by nerve to quadratus femoris, superior by nerve to obturator internus.

Quadratus femoris
Origin: From outer border of tuberosity of ischium; fibres run horizontally outwards.

Insertion: Quadrate tubercle on upper part of inter-trochanteric crest of femur.

Nerve supply: Nerve to quadratus femoris (L. 4, 5, S. 1).

The hamstring group of muscles

These long muscles of the back of the thigh flex the leg at the knee and extend the thigh at the hip. They are the semitendinosus, semimembranosus and biceps femoris and (except for the short head of biceps) they arise from the ischial tuberosity. They are supplied by the sciatic nerve— the short head of biceps by the common peroneal (L. 5, S. 1), the others by the tibial nerve (L. 5, S. 1).

The *semimembranosus* arises from the upper lateral area on the back of the tuberosity of the ischium; the *semitendinosus* and *long head of biceps* together from the lower medial area. The upper part of semimembranosus consists of a wide, thin tendon, which passes downwards deep to the other two.

The long head of biceps passes laterally and downwards across the semimembranosus tendon to join its short head in the lower part of the thigh. This *short head of the biceps* arises from the linea aspera and upper part of the lateral supracondylar ridge. The two heads join to form a tendon which is inserted into the lateral aspect of the head of the fibula inferior to the fibular collateral ligament of the knee.

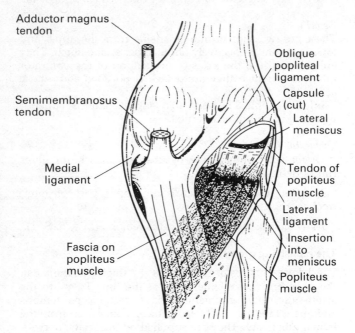

Fig. 33 Dissection behind right knee to show insertion of popliteus into lateral meniscus.

The semimembranosus and semitendinosus run vertically down the medial aspect of the back of the thigh, with the latter superficial.

Insertion of semimembranosus: Primarily into a groove on back of medial condyle of tibis; extensions go to oblique line of tibia as fascia over popliteus, to back of capsule of knee joint (oblique popliteal ligament), to bone inferior to medial ligament (Fig. 33).

Insertion of semitendinosus: Medial side of upper part of shaft of tibia behind sartorius and inferior to gracilis.

Popliteal fossa is diamond-shaped hollow at back of knee (compare cubital fossa on front of elbow) bounded above by diverging hamstrings and below by converging heads of gastrocnemius. Floor is formed by lower posterior surface of femur, capsule of knee joint and popliteus. Passing vertically through it, about its middle are the tibial nerve and popliteal vessels and laterally common peroneal nerve deep to tendon of biceps.

MUSCLES OF THE LEG

These consist of three groups: anterior or extensor, supplied by deep peroneal nerve; posterior or flexor, supplied by tibial nerve; and lateral or evertor supplied by superficial peroneal nerve.

Note: Owing to the fact that the fibula is in a plane posterior to the tibia the anterior compartment faces laterally as well as anteriorly.

The *deep fascia of the leg* forms a sheath for all these muscles and sends septa in between them; it is attached to anterior border of tibia, and sweeps round front, lateral side and the back of leg to reach tibia again at its posteromedial border.

At ankle deep fascia forms retinacula like those at the wrist to hold the various tendons in place.

The anterior or extensor muscles (anterior compartment)

Tibialis anterior
Origin: Upper two-thirds of lateral surface of tibia, and interosseous membrane.

Insertion: Medial and inferior surface of medial cuneiform and base of 1st metatarsal.

Action: Dorsiflexion (extension) and inversion of foot.

Extensor hallucis longus
   Origin: Medial part of middle third of anterior surface of fibula and interosseous membrane.

Insertion: Dorsal aspect of base of terminal phalanx of great toe.

Action: Extension of great toe and dorsiflexion (extension) of foot.

Extensor digitorum longus
   Origin: Upper two-thirds of anterior surface of fibula, lateral to extensor hallucis longus, and interosseous membrane.

Insertion: By four tendons into the four lateral toes; each tendon is inserted into the bases of the middle and distal phalanges, as in fingers (p. 87).
   Each tendon to the 2nd, 3rd and 4th toes is reinforced by a tendon of extensor digitorum brevis (p. 112).

Action: Extension of toes and dorsiflexion of foot.

Peroneus tertius
Not really a peroneal muscle, but a part of extensor digitorum longus.

Origin: Lower part of anterior surface of shaft of fibula in continuity with extensor digitorum longus.

Insertion: Dorsal aspect of base and shaft of 5th metatarsal.

Nerve supply: Note that this is the deep peroneal, the same as the other extensors.

Action: Dorsiflexion (extension) and eversion of foot.

Important relations of the anterior compartment muscles: The anterior tibial artery and deep peroneal nerve lie on the interosseous membrane, first between tibialis anterior and extensor digitorum longus, and then between tibialis anterior and extensor hallucis longus; they pass deep to the tendon of extensor hallucis longus at the level of the ankle joint.

Extensor retinacula
Two bands, one above, the other in front of and below the ankle.
Superior retinaculum extends from upper part of fibular malleolus to shaft of tibia just above malleolus. Pierced by tibialis anterior tendon, other tendons pass deep to it.
Inferior retinaculum (Y-shaped) extends from lateral part of calcaneus medially and splits; upper limb goes to medial malleolus, lower to plantar aponeurosis. Pierced by all the tendons.
The tendons pass downwards deep to these bands. Tibialis anterior has a synovial sheath extending deep to both retinacula; extensor hallucis has a separate synovial sheath and extensor digitorum longus and peroneus tertius have a common sheath deep to the lower band only.

The posterior or flexor muscles of the calf

These consist of superficial deep groups.
The superficial group (gastrocnemius, plantaris and soleus) join to form the stout tendo calcaneus, which is inserted into the back of the calcaneus. The gastrocnemius and plantaris cross the back of the knee joint and therefore flex the leg at the knee as well as the foot at the ankle.

THE SUPERFICIAL MUSCLES OF THE CALF

The *gastrocnemius* arises by two heads from the femur.

Origin—medial head: From the back of the femur immediately above the medial condyle; *lateral head:* Upper part of lateral aspect of lateral condyle (note asymmetry of these two heads).

Insertion: Muscle fibres are replaced about halfway down the leg by *tendo calcaneus (Achillis),* which is inserted into middle third of back of calcaneus. A bursa separates tendon from upper third of back of calcaneus.

Action: Flexion of foot at ankle and of leg at knee.

Soleus
Origin: Upper fourth of posterior aspect of fibula; soleal (oblique popliteal) line of tibia and middle third of medial border of shaft of tibia; fibrous arch over popliteal vessels.

Insertion: Deep aspect of tendo calcaneus, and thence to calcaneus.

Action: Flexor of foot at ankle. Powerful multipennate muscle.

Plantaris
A muscle with a 6 cm muscle belly and a very long tendon.

Origin: Lower part of lateral supracondylar ridge. Forms a tendon passing downwards and medially between soleus and gastrocnemius.

Insertion: Blends with medial part of tendo calcaneus.

Action: Accessory to gastrocnemius; the muscle in man is a rudiment of a large muscle of the toes in certain animals.
Note: The superficial calf muscles, especially the soleus, have a special postural function in standing. They maintain the balance of the body at the ankle, since the body is falling forwards in the upright position. The calf muscles are important in raising the body on to the toes in walking and running, before the foot leaves the ground.

THE DEEP MUSCLES OF THE CALF

These are a short muscle of the knee (popliteus) and three muscles inserted into the foot.

Popliteus
  Origin: Lateral half of muscle by a round tendon from anterior end of popliteal groove on lateral femoral condyle. Medial half of muscle by a flat aponeurosis attached to lateral meniscus and capsule of joint (Fig. 33).

  Insertion: Popliteal surface of back of tibia above soleal line.

  Nerve supply: From tibial by a branch curving around lower border of muscle and entering deep surface.

  Action: Rotates femur laterally on tibia or tibia medially on femur. Also controls position of lateral meniscus, protecting it from being crushed between condyles of femur and tibia. Not a flexor of knee.

Flexor hallucis longus
  Origin: Lower three-quarters of posterior surface of fibula and the interosseous membrane. Large powerful multipennate muscle.

  Insertion: Plantar surface of base of terminal phalanx of big toe.

  Nerve supply: Tibial (S. 1, 2).

  Action: Flexes big toe. Important in maintaining medial longitudinal arch of foot (p. 43).
  Note: The tendon is very deep at ankle; grooves the talus and the inferior surface of the sustentaculum tali; runs forwards in the sole to cross deep to tendon of flexor digitorum longus to which it gives a slip.

Flexor digitorum longus
  Origin: The medial half of the back of the tibia below the oblique soleal line. It is a bipennate muscle.

  Insertion: Into the four outer toes. Each tendon is inserted into the base of the distal phalanx after passing through the split flexor digitorum brevis tendon; identical with arrangement in hand.

Action: Flexion of the toes.

Nerve supply: Tibial nerve (S. 1, 2).
Note: Both long flexors are markedly active when standing on tiptoe and at the toe-off stage of the step in walking and running.

Tibialis posterior
   Origin: Lateral half of back of tibia below soleal line; upper two-thirds of medial surface of fibula; interosseous membrane.

   Insertion: Primarily into the tuberosity of the navicular. From this insertion slips pass backwards to the sustentaculum tali, laterally to cuboid, intermediate and lateral cuneiforms and bases of 2nd, 3rd and 4th metatarsals and forwards to medial cuneiform and base of 1st metatarsal.

   Action: Flexion and inversion of the foot. With the foot on the ground, the muscle is an important support of the longitudinal arches of the foot. Its tendon lies directly alongside the spring ligament.

   Nerve supply: Tibial nerve (L. 4).
   Note: The posterior tibial vessels and tibial nerve lie superficial to tibialis posterior, between flexors hallucis longus and digitorum longus.

The *flexor retinaculum* extends from the tibial malleolus to the medial margin of the back of the calcaneus; it covers the three deep tendons (which lie in synovial sheaths) and the posterior tibial vessels and tibial nerve (from medial to lateral—tibialis posterior, flexor digitorum longus, vessels, nerve, flexor hallucis longus).

The peroneal or evertor group

Two muscles on the lateral aspect of the leg—peroneus longus and brevis; both arise from the lateral surface of the fibula.

Peroneus longus
   Origin: Upper two-thirds of lateral aspect of shaft of

fibula. Tendon lies on peroneus brevis behind lateral malleolus, crosses calcaneofibular ligament, passes below peroneal trochlea (tubercle), lies in groove of cuboid and crosses sole of foot obliquely. Sesamoid fibrocartilage or bone in tendon where it turns into sole at cuboid.

Insertion: Lateral aspect of medial cuneiform and base of 1st metatarsal.

Nerve supply: Superficial peroneal nerve (L. 5, S. 1), which pierces upper part of muscle.

Action: Eversion and weak plantar flexion of foot; helps in supporting the lateral longitudinal and transverse arches of the foot.

Peroneus brevis: Anterior to longus.

Origin: Lower two-thirds of lateral aspect of fibula, overlapping lower part of peroneus longus, which lies behind it.

Insertion: Styloid process of base of 5th metatarsal. Tendon lies in front of peroneus longus on back of lateral malleolus and above peroneal tubercle on lateral aspect of calcaneus.

Nerve supply: Superficial peroneal (L. 5, S. 1).

Action: Eversion and weak plantar flexion of foot.

Peroneal retinacula and synovial sheaths
Peroneal tendons are invested by common sheath on back of lateral malleolus where retained by *superior retinaculum* attached to fibular malleolus and calcaneus. As they separate to take up their respective positions above and below the peroneal tubercle (Fig. 17), each carries its own sheath forwards from common sheath. Small loops of fibrous tissue above and below form *inferior retinaculum* for the tendons with their sheaths.

MUSCLES OF THE FOOT

These are short muscles whose origin and insertion are entirely within the foot.

Dorsum of the foot

Extensor digitorum brevis

   Origin: From anterior part of the upper surface of calcaneus; forms a fleshy mass partly under cover of the inferior extensor retinaculum.

   Insertion: By four tendons into the four medial toes. Tendon to great toe (*extensor hallucis brevis*) is inserted independently into dorsal surface of base of proximal phalanx; those to 2nd, 3rd and 4th toes join common extensor expansion.

   Nerve supply: Lateral terminal branch of deep peroneal on its deep surface.

Sole of the foot

These are best described in four layers from the surface inwards; the most superficial structure in the sole is the plantar aponeurosis, a thickened, deep fascia, closely resembling the similar layer in the palm (p. 91).

   The *plantar aponeurosis* consists of a thick, strong central part, which is attached posteriorly to the medial process of the calcanean tuberosity, and thinner parts on each side which form the ordinary deep fascia covering the muscles. The central portion spreads out as it passes forwards, and splits into five digitations, one to each toe, which blend with the fibrous sheaths of the flexor tendons of the toes (cf. the palmar aponeurosis).

FIRST LAYER OF MUSCLES

These are the *abductor hallucis*, *abductor digiti minimi* and *flexor digitorum brevis*.

Origin: The three muscles arise from front of the tubercle of calcaneus. Abductor hallucis is medially placed; abductor digiti minimi arises from lateral and medial processes of the tubercle; the flexor digitorum brevis arises from the medial process only, superficial to the abductor digiti minimi.

Insertions: Abductor hallucis into medial side of base of proximal phalanx with medial tendon of flexor hallucis brevis in which there is a sesamoid bone; abductor digiti minimi into lateral side of base of proximal phalanx of little toe; flexor digitorum brevis tendons split over the flexor digitorum longus tendons to go to the margins of the middle phalanx of each of the four lateral toes.

Nerve supply:

Abductor hallucis ⎱	medial plantar nerve
Flexor digitorum brevis ⎰	(S. 1, 2)
Abductor digiti minimi	lateral plantar nerve
	(S. 1, 2)

Action: As named. Help to maintain longitudinal arches.

SECOND LAYER OF MUSCLES

These are flexor digitorum accessorius and lumbricals, both being associated with the tendons of the flexor digitorum longus, which together with the tendon of the flexor hallucis longus, lie in this layer immediately deep to flexor digitorum brevis.

Flexor digitorum accessorius
Origin: By two heads, medial head (muscular) from hollow surface of calcaneus under sustentaculum tali and lateral head (smaller and tendinous) from inferior surface of calcaneus in front of tubercle and from long plantar ligament.

Insertion: Deep surface of tendons of flexor digitorum longus.

Nerve supply: Lateral plantar (S. 1, 2).

Action: Flexes terminal phalanges when belly of long flexor is shortened in plantar flexion of foot. May also straighten oblique line of pull of long flexor tendons.

Lumbricals
Four very slender muscles arising from adjacent sides of tendons of flexor digitorum longus, except first which is from medial side of first tendon.

Insertion: Into extensor expansions on dorsum of lateral four toes passing to medial side of each toe.

Nerve supply: 1st by medial plantar; 2nd, 3rd and 4th (bicipital) by lateral plantar.

Action: Prevent terminal phalanges from buckling due to pull of flexor digitorum longus—they keep the toes straight.

THIRD LAYER OF MUSCLES

Flexor hallucis brevis
Origin: Tendon of tibialis posterior; by oblique tendinous slip from inferior surface of cuboid.

Insertion: By two slips to medial and lateral sesamoid bones of great toe, and so to sides of proximal phalanx.

Nerve supply: Medial plantar nerve.

Adductor hallucis
Origin: By two heads, *transverse* from ligaments of plantar aspect of metatarsophalangeal joints of 3rd, 4th and 5th toes and *oblique* from sheath of peroneus longus and bases of 2nd, 3rd and 4th metatarsals.

Insertion: Through lateral sesamoid into lateral aspect of base of proximal phalanx of great toe with lateral tendon of flexor hallucis brevis.

Nerve supply: Lateral plantar.

Flexor digiti minimi brevis
   Origin: Base of 5th metatarsal and sheath of peroneus longus.

   Insertion: Lateral aspect of base of proximal phalanx.

   Nerve supply: Lateral plantar.

FOURTH LAYER OF MUSCLES

These are three plantar and four dorsal interossei. They resemble those of the hand but the movements they produce are related to the second toe as compared with the third digit in the hand.

The four *dorsal interossei* abduct from line of second toe, and therefore the first and second act on the second toe.

   Origin: From the adjacent sides of the metatarsals in each intermetatarsal space.

   Insertion: First—medial side of base of proximal phalanx of 2nd toe; second—lateral side of base of proximal phalanx of 2nd toe; third—lateral side of base of proximal phalanx of 3rd toe; fourth—lateral side of base of proximal phalanx of 4th toe.

The three *plantar interossei* adduct, and therefore none is inserted into the 2nd toe.

   Origin: Medial side of 3rd, 4th and 5th metatarsals.

   Insertion: Medial side of bases of proximal phalanges of same toes.

   Nerve supply: Lateral plantar (deep division to medial three spaces, superficial division to fourth space).

THE MUSCLES OF THE ABDOMINAL WALL

The abdominal wall (Fig. 34) consists principally of three sheets of muscle which are fleshy at the sides but

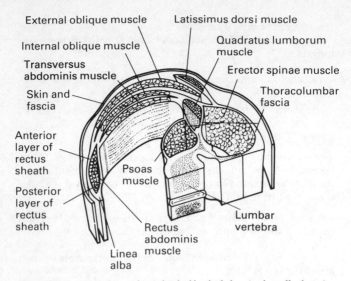

Fig. 34 Section through right half of abdominal wall showing thoracolumbar fascia and rectus sheath.

aponeurotic in front and behind; they are the external oblique, internal oblique and transversus from without inwards. In addition, there is a vertical muscle on either side of the midline in front—rectus abdominis. As the aponeuroses pass forwards they ensheath the rectus; deep to the lower part of this sheath above the pubis is a small muscle, pyramidalis. Posteriorly the abdominal wall is completed by the quadratus lumborum, between the last rib and the iliac crest. The psoas (p. 94) also forms part of the posterior abdominal wall near the vertebral column.

Actions of the abdominal muscles
They retain the viscera within the abdominal cavity and by their tone they may maintain their position. They can compress hollow viscera and assist in expulsive efforts—defaecation, micturition, parturition, vomiting. In addition they are involved in movements of the vertebral column, especially the recti (e.g. they flex the trunk when lying

supine and control extension in the upright position; the obliques are involved in rotatory movements). The obliques contract in deep expiration to push up the diaphragm, and in coughing.

External oblique

Origin: Outer surfaces of lower eight ribs by slips, the upper of which interdigitate with serratus anterior and the lower with latissimus dorsi.

Insertion: Most posterior fibres pass downwards to anterior half of iliac crest; rest of muscle, directed downwards and forwards, ends in an aponeurosis which is attached (a) to the anterior superior iliac spine and the pubic tubercle and crest (between these two points it is folded inwards and thickened to form the *inguinal ligament* to which is attached the deep fascia of the thigh), and (b) to its fellow of the opposite side in the midline forming the fibrous *linea alba*, which extends from xiphisternum above to symphysis pubis below.

Nerve supply: Lower six intercostal nerves.

Note: The linea alba is formed not only by the external oblique aponeurosis, but by the aponeuroses of the internal oblique and transversus, all three blending in the midline.

The *inguinal (Poupart's) ligament* bridges the femoral sheath, femoral nerve, psoas and iliacus as these pass from the abdomen into the thigh. A deep extension of the medial end of the ligament to the pectineal line is called the *lacunar ligament*. The lateral free edge of the lacunar ligament continues on to the pubic bone as the *pectineal ligament* and forms the medial boundary of the femoral ring (the upper end of the femoral canal). The pectineal ligament forms its posterior boundary.

The *superficial inguinal ring* is a triangular opening in the external oblique aponeurosis situated above the pubic crest. It is the external opening of the inguinal canal, and transmits the spermatic cord (male), round ligament of uterus (female) and ilio-inguinal nerve.

Note three free borders of external oblique: (1) posterior,

fleshy, a boundary of lumbar triangle, (2) inferior, aponeurotic (inguinal ligament), (3) superior, upper border of rectus sheath, from 5th rib to sternum. Pectoralis major arises from this aponeurosis.

The *lumbar triangle* is bounded by iliac crest, external oblique laterally and latissimus dorsi medially. Floor formed by internal oblique.

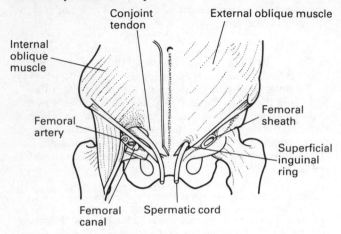

Fig. 35 Inguinal canal.

Internal oblique (Figs. 34 and 35) (and *cremaster*)

Origin: From thoracolumbar fascia attached to lumbar vertebrae; from anterior two-thirds of iliac crest and lateral two-thirds of inguinal ligament. Fibres are directed forwards and upwards.

Insertion: By muscular fibres into lower six costal cartilages; by aponeurotic sheet which blends in midline with muscle of opposite side (linea alba); most inferior fibres from inguinal ligament arch over spermatic cord and descend to be attached to pubic crest and pectineal line forming *conjoined tendon*, common to it and transversus. In male, lowest fibres form loops around spermatic cord and constitute separate supporting muscle, the *cremaster*.

Nerve supply: Lower six intercostal and first lumbar (iliohypogastric and ilio-inguinal); cremaster by genital branch of genitofemoral nerve (L. 2).

Transversus abdominis

Origin: From deep surface of lower six costal cartilages interdigitating with diaphragm; posteriorly from thoraco-lumbar fascia attached to lumbar vertebrae; from anterior two-thirds of medial lip of iliac crest and lateral half of inguinal ligament. Fibres run horizontally forwards.

Insertion: By aponeurotic sheet joining its fellow of opposite side in midline (linea alba); below to pubic crest and pectineal line by conjoined tendon common to it and internal oblique.

Nerve supply: Lower six intercostal and iliohypogastric and ilio-inguinal.

Neurovascular plane

Nerves and vessels lie in abdominal wall on deep surface of internal oblique, between it and transversus and pass forwards to rectus.

Note: Transversalis fascia is a thin layer lining inner surface of transversus muscle. It is continuous with the general layer of fascia lining the posterior abdominal wall. In the inguinal region it becomes continuous across the back of the inguinal ligament with the iliac fascia (p. 95); where the femoral vessels pass into the thigh they carry the transversalis fascia with them as the anterior layer of the femoral sheath and the iliac fascia as the posterior layer.

Inguinal canal

Canal in abdominal wall above medial half of inguinal ligament, about 5 cm long. Transmits spermatic cord (male), round ligament (female).

Boundaries (Fig. 35): *Posterior wall*—transversalis fascia laterally and conjoined tendon medially; *Anterior wall*—external oblique aponeurosis, reinforced in lateral part by origin of internal oblique from inguinal ligament; *Roof*—fibres of internal oblique and transversus arching over cord to form conjoined tendon; *Floor*—inverted edge of inguinal ligament and medially, lacunar ligament (almost horizontal in the upright position).

Entrance: Circular *deep inguinal ring* in transversalis fascia about 1 cm diameter. Medial side of ring reinforced by *interfoveolar ligament*, looping downwards from transversus muscle to inguinal ligament. Ring 1 cm above midpoint of inguinal ligament; inferior epigastric artery passes upwards medial to its medial edge.

Exit: Superficial inguinal ring in external oblique aponeurosis. Triangular with base medial to pubic tubercle and sides (*crura*) sloping upwards and laterally from pubic tubercle and symphysis. Apex reinforced by *intercrural fibres* at right angles to those of external oblique aponeurosis. *Reflected part of inguinal ligament* passes from lateral crus upwards and medially behind external oblique and superficial to conjoint tendon to linea alba.

Rectus abdominis

Origin: Medial head, tendinous, from front of body of pubis close to symphysis; lateral head, muscular, from crest of pubis.

Insertion: Anterior aspect of 5th, 6th and 7th costal cartilages.
Note: The two recti diverge a little at their upper ends, and become thinner and wider than below.

Nerve supply: Lower six intercostals.

Action: The rectus flexes the pelvis on the trunk, or the trunk on the pelvis.

Rectus sheath and tendinous intersections

As the aponeurotic sheets of the two obliques and transversus pass medially to the midline they form the sheath of the rectus. The internal oblique aponeurosis splits at the lateral edge of the rectus into two laminae, one in front of the rectus and one behind; the transversus aponeurosis fuses with the posterior layer, and the external oblique aponeurosis with the anterior. Over the lower fourth of the rectus, however, all three aponeuroses pass in front, thus leaving the sheath deficient posteriorly. The lower crescentic free margin is called the *arcuate line (semilunar fold of Douglas)*. The fused aponeuroses in the midline between the recti constitute the *linea alba*.

The superior epigastric artery, from internal thoracic, runs downwards from thorax between posterior layer of sheath and muscle and anastomoses with inferior epigastric artery, from the external iliac, running upwards behind rectus muscle and then between posterior layer and the muscle. The lower intercostal nerves run horizontally behind the muscle, supply it, and end by piercing the muscle as an anterior cutaneous branch. If entering abdomen parasagittally, rectus muscle is split (not cut) longitudinally or cut transversely or pulled to one side.

The anterior layer of the sheath is attached to the muscle by irregular transverse tendinous intersections, which dip at intervals into the substance of the rectus but do not reach the posterior layer. These are usually three in number: one about the level of the umbilicus, one at the xiphisternum, and one between these. The space within the sheath in front of the rectus thus consists of several compartments, but the space behind the rectus is long and undivided.

Pyramidalis
Small triangular muscle inside rectus sheath, in front of lower part of rectus arises from crest of pubis and is inserted into the linea alba. Supplied by subcostal nerve (T. 12) and may tighten linea alba.

Quadratus lumborum
Origin: Posterior quarter of iliac crest: iliolumbar ligament; tips of transverse processes of lumbar vertebrae.

Insertion: Medial half of inferior border of 12th rib.

Nerve supply: 2nd and 3rd lumbar straight from ventral rami.

Action: Lateral flexion of vertebral column; fixator of last rib for action of diaphragm during inspiration.

Lies in anterior compartment of thoracolumbar fascia (p. 67).

THE MUSCLES OF THE CHEST WALL

Morphology of muscles

The three muscular layers seen in abdominal wall (p. 115) can be recognized in thorax: (1) external oblique represented by external intercostal, (2) internal oblique represented by internal intercostal, (3) transversus abdominis represented by a number of muscles (intercostalis intimi etc., p. 123).

Note: Neurovascular plane lies between (2) and (3) in both thorax and abdomen; all muscles are supplied by thoracic spinal nerves (Fig. 36).

In ventral midline in abdomen three layers fuse into longitudinal rectus abdominis. (Note segmental supply from lower intercostal nerves.)

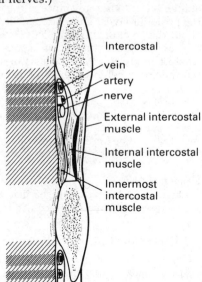

Intercostal
vein
artery
nerve
External intercostal muscle
Internal intercostal muscle
Innermost intercostal muscle

Fig. 36 Three intercostal muscles and the neurovascular plane.

External intercostals
  Origin: Outer border of costal groove of rib above.

  Insertion: Fibres directed downwards and forwards to upper border of rib below.

  Extent of muscle: From superior costotransverse ligament posteriorly to costochondral junction; then replaced to sternum by *external (anterior) intercostal membrane.*

  Nerve supply: Corresponding intercostal nerves by collateral branches.

  Action: May elevate ribs in inspiration.

Internal intercostals
  Origin: Floor of costal groove of rib above.

  Insertion: Fibres directed downwards and backwards to upper border of rib below.

  Extent of muscle: From side of sternum to angle of rib, then replaced to superior costotransverse ligament by *internal (posterior) intercostal membrane.*

  Nerve supply: Corresponding intercostal nerves by collateral branches.

  Action: Anteriorly, may raise ribs in inspiration; more posteriorly may depress ribs in expiration.

Intercostales intimi
  Origin: Medial border of costal groove of rib above (membranous).

  Insertion: May cross one or two ribs; blend with internal intercostal at upper border of a rib.

  Nerve supply: Corresponding intercostal nerves.

  Action: As internal intercostals.

Transversus thoracis (sternocostalis)

Origin: Deep aspect of xiphoid process, lower part of body of sternum and 5th, 6th and 7th costal cartilages.

Insertion: Fibres radiate upwards and laterally to backs of 2nd to 6th costal cartilages.

Nerve supply: Intercostal nerves.

Action: May raise xiphisternum (i.e. weak inspiratory effect).

Subcostals
Ill-defined slips extending over several ribs and lining posterolateral part of lower chest wall. Supplied by intercostal nerves.

Levatores costarum
Eleven small muscles, each arising from the tip of a thoracic transverse process and inserted into upper border of rib below, near angle.

Nerve supply: Dorsal rami of corresponding intercostal nerves.

THE DIAPHRAGM

This is a domed musculotendinous sheet separating the thorax from the abdomen. It consists of a central trilobed tendon to which are attached peripheral skeletal muscle fibres. These fibres complete the septum between thorax and abdomen, and are attached all round the body wall; they are described as taking origin from the body wall and being inserted into the margin of the central tendon.

Origin: (1) Posterior fibres (a) by right and left crura, from fronts of upper lumbar vertebrae (three on right side, two on left); (b) from the medial and lateral arcuate ligaments (Fig. 78), the upper borders of the psoas fascia and quadratus lumborum fascia respectively. (2) Lateral fibres, from the inner surfaces of the lower six costal cartilages, interdigitating with the transversus abdominis (p. 119). (3) Anterior fibres, small slips from back of xiphisternum.

Insertion: Each part arches upwards and inwards to be inserted into the adjacent part of the central tendon.

Nerve supply: Phrenic (C. 3, 4, 5, mainly 4); peripheral part by lower six intercostal nerves (sensory); crura by phrenic nerves.

Action: (1) Principal muscle of inspiration; the muscular fibres depress the dome and so increase the vertical diameter of the chest. The resulting decrease in intrathoracic pressure facilitates the entry of air into the lungs. The abdominal viscera are pushed downwards and the anterior abdominal wall forwards. (2) Remains contracted when anterior abdominal muscles contract and increase intra-abdominal pressure in expulsive efforts, e.g. defaecation, micturition, labour. Pelvic diaphragm relaxes. In vomiting and coughing reverse occurs—pelvic diaphragm contracts and diaphragm (thoracic) relaxes.

Openings in diaphragm: (1) For inferior vena cava (also transmits right phrenic nerve) 2 cm to right of midline in right leaf of central tendon at level of 8th thoracic vertebra. Pulled open when diaphragm contracts, aiding venous return. (2) For oesophagus, through right crus as it passes to the left and crosses the left crus; 2—3 cm to left of midline at level of 10th thoracic vertebra. Also transmits gastric nerves (i.e. vagi) and oesophageal vessels (artery, veins, lymphatics) from left gastric vessels. Closed by muscular fibres when diaphragm contracts. (3) For aorta between tendinous margins of crura at level of 12th thoracic vertebra. Does not pass through diaphragm but behind it, and is unaffected when diaphragm contracts. Also transmits vena azygos and thoracic duct.

Splanchnic nerves pass through crura, sympathetic trunk behind medial arcuate ligament. Lower intercostal nerves pass between digitations into neurovascular plane of anterior abdominal wall. Subcostal nerve behind lateral arcuate ligament. Left phrenic nerve pierces left dome.

THE MUSCLES OF THE PELVIS

Two of these muscles, levator ani and coccygeus, are limited to the pelvis. Two others line pelvic wall, piriformis on front of sacrum and obturator internus on the lateral wall of the true pelvis; these two muscles, however, act on the femur, and are described with the muscles of the lower limb (p. 102).

Levator ani muscles
From side wall of pelvis as thin sheets which meet in midline and close greater part of outlet of the pelvis (*pelvic diaphragm*).

Origin: (1) Back of body of pubis. (2) Spine of ischium. (3) Between these from fascia covering obturator internus along a thickening between these two points (*white line*).

Insertion: Into midline, extending from side of coccyx, overlapping pelvic surface of coccygeus, into anococcygeal body (between coccyx and anal canal), round anal canal, into central tendon of perineum (perineal body), round vagina. Some fibres from pubis form slings—*puborectalis, pubovaginalis, pubo-urethralis*. Fibres inserted into prostate are called *levator prostatae*.

Nerve supply: S. 4 on pelvic surface and inferior rectal and perineal on perineal surface (from pudendal).

Action: (1) Muscles as a whole form diaphragm of pelvic outlet, supporting pelvic viscera and resisting downward pressure of abdominal muscles if necessary. Relaxation occurs if emptying hollow viscus downwards. (2) They form special sphincters of anal canal and vagina. (3) Some fibres into urethra, vagina and upper part of anal canal may pull these structures open and lead to the rapid fall of pressure in anal canal in defaecation and in urethra in micturition.

Coccygeus
A small triangular muscle behind, and in the same plane as the levator ani.

Origin: Spine of ischium.

Insertion: Side of coccyx and lowest part of sacrum.

Nerve supply: S. 4.

Action: The two muscles pull the coccyx forward and form the posterior part of the pelvic diaphragm.

Pelvic fascia
Parietal pelvic fascia is a strong membrane covering the muscles of the pelvic walls; it is attached to bone at margins of muscles.

Visceral pelvic fascia is loose and cellular over movable (levator ani) or distensible (bladder and rectum) structures; strong and membranous over fixed or non-distensible structures (prostate, base of bladder). The rectovesical fascia lies between the bladder and rectum.

THE MUSCLES OF THE PERINEUM

Note: Anatomically the perineum is the region bounded by the symphysis pubis in front, the coccyx behind and the ischial tuberosities laterally. A transverse line between the tuberosities divides the region into a posterior *anal triangle* and anterior *urogenital triangle*. Clinically the term *perineum* usually refers to the urogenital triangle although in obstetrics it refers to the perineal body between the anal and vaginal openings.

MUSCLES OF THE ANAL CANAL

Anal canal
3 cm long, extending from anorectal junction at levator ani (puborectalis) to anocutaneous junction at exterior. Has two sphincters.

Internal sphincter: Smooth muscle, surrounds upper two-thirds of anal canal; a thickening of circular coat of rectum, supplied by sympathetic nerves (pelvic plexus). Incompetent alone, but assists external sphincter.

External sphincter: Skeletal muscle, surrounds lower two-thirds of anal canal; consists of three rings that make a tube of muscle. Deepest ring fused posteriorly with puborectalis, called *deep part* of sphincter; nerve is inferior rectal. Middle ring is *superficial* attached behind to coccyx, in front to central tendon of perineum; nerve is perineal branch of S. 4. Third ring, *subcutaneous*, is unattached, lies deep to peri-anal skin; supplied by inferior rectal nerve. Most important part is anorectal ring, formed by deep part and puborectalis. Is essential for continence of faeces and flatus.

Ischiorectal fossa
Space between anal canal and side wall of pelvis lined by obturator internus and its fascia which is condensed around internal pudendal vessels and pudendal nerve to form *pudendal canal*. Roof is inferior (perineal) surface of levator ani; hence fossa has prolongation forwards superior to sphincter urethrae on each side of membranous urethra. The space contains loose fat; infection here causes *ischiorectal abscess*, which may burst into anal canal.

MUSCLES OF THE UROGENITAL TRIANGLE

Muscles of the superficial perineal pouch

This pouch is bounded inferiorly by a membranous layer of superficial fascia (Colles' fascia) and superiorly by the perineal membrane (inferior layer of urogenital dia-phragm).

Bulbospongiosus
In the male a thin sheet covering the corpus spongiosum.

Origin: Central tendon of perineum and ventral midline raphe.

Insertion: Into aponeurosis on dorsum of penis, some fibres encircling corpus spongiosum only, and others encircling the corpora cavernosa as well.

Action: Compresses bulbous part of urethra at end of micturition. Empties urethra.

In the female the bulb of corpus spongiosum is split and encircles the vagina so that the bulbospongiosus also encircles the vagina and anteriorly the corpus spongiosum of the clitoris.

Action: May act as sphincter of the orifice of the vagina.

Ischiocavernosus
Covers the crus of the corpus cavernosum on each side.

Origin: Medial aspect of ischiopubic ramus.

Insertion: Sides of crus.

Action: Compresses crus during erection.

The same muscle in the female covers the crus clitoridis.

Superficial transverse perinei
Origin: Medial surface of tuberosity of ischium, behind ischiocavernosus.

Insertion: Central tendon of perineum.

Action: Supports central part of perineum.

Muscles of deep perineal pouch

This pouch is bounded inferiorly by the perineal membrane and superiorly by the superior layer of the urogenital diaphragm.

Sphincter urethrae (external urethral sphincter)
Origin: Edge of pubic arch in region of junction of pubis and ischium.

Insertion: Raphe in front of and behind urethra; some inner fibres are not attached to bone and encircle urethra.

Action: Can maintain urinary continence in absence of vesical sphincter.

Deep transverse perinei

  Origin: Ischial tuberosity behind sphincter urethrae.

  Insertion: Central tendon of perineum.

  Action: Supports central part of perineum.

  Nerve supply of all these muscles is perineal branch of pudendal nerve (S. 2, 3, 4).

  Note: All three layers of the pouches fuse together posteriorly behind the transverse perineal muscles and are attached laterally to ischiopubic rami. Anteriorly the superficial layer (Colles') extends round the external genitalia forwards into abdominal wall as membranous layer of superficial fascia. The two layers of urogenital diaphragm do not reach symphysis pubis leaving a gap through which deep dorsal vein of penis (or clitoris) passes into pelvis. The *urogenital diaphragm* includes the fascial layers and muscles of the deep pouch.

3
The Heart and Arteries

THE HEART

The heart is a hollow muscular organ enclosed in the pericardium. It has an irregular conical shape and is attached at its base to the great blood vessels. It is otherwise free within the pericardial sac.

Position
The heart is placed obliquely. The base is directed upwards, backwards and to the right, and the apex downwards, forwards and to the left.

Surface marking of the outline of the heart (Fig. 37)

Superior: Line from lower border of 2nd left to upper border of 3rd right costal cartilage.

Inferior: Line from 6th right sternocostal articulation to apex.

Right side: Line drawn nearly vertically 3 cm from midline of sternum from 3rd to 6th costal cartilages.

Left side: Line from lower border of 2nd left costal cartilage 1 cm from sternum, downwards and to the left to apex.

The apex: Corresponds to a point in the 5th left intercostal space 8 cm from the midline.

Surface marking of position of the valves (Fig. 37)

Pulmonary: Behind upper border of 3rd left costal cartilage near sternum.

Aortic: Behind left half of sternum, level with lower border of 3rd costal cartilage.

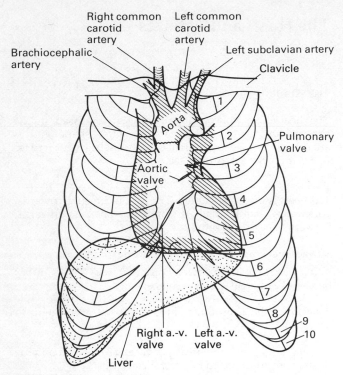

Fig. 37 Surface markings of outline of heart, aorta and its large branches, and liver.

Right atrioventricular (tricuspid): Extends obliquely from midline at level of 4th costal cartilage down to junction of 6th cartilage with sternum.

Left atrioventricular (mitral): At level of junction of 4th left costal cartilage with sternum.

Divisions

The heart is divided longitudinally into right and left halves by a vertical septum which almost lies in the coronal plane.

Atria are divided from *ventricles* by *coronary (atrioventricular) sulcus* which lies approximately in the sagittal plane. Right atrium lies more or less in front of left atrium and right ventricle in front of left ventricle.

Atrioventricular fibrous ring separates muscle of atrium from that of ventricle. Two rings joined (right in front of left) to make figure of 8; membranous part of interventricular septum is attached to line of junction.

Circulation of blood through the heart

The right atrium receives venous blood from the venae cavae and coronary sinus; thence blood passes into the right ventricle, whence it is conveyed to the lungs by the pulmonary trunk. After passing through the lungs, the

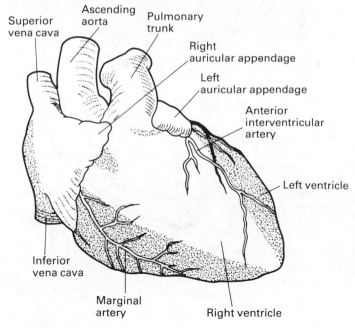

Fig. 38 Anterior surface of heart.

blood enters the left atrium by the pulmonary veins; from there it is conveyed to the left ventricle, and from there to the aorta, which with its branches distributes it to the body.

External appearance of the heart

Anteriorly most of the surface is formed by the right ventricle with the right atrium to the right, forming the right border, and the left ventricle to the left forming the left border and apex (Fig. 38). Most of the left ventricle is on the inferior (diaphragmatic) surface of the heart. A small part of the right ventricle is on this surface (Fig. 39). Posteriorly the left ventricle is to the left, the left atrium is somewhat superior and the right atrium is on the right (Fig. 40).

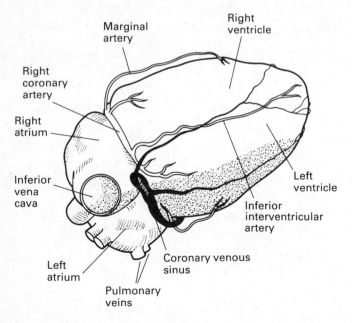

Fig. 39 Diaphragmatic surface of heart.

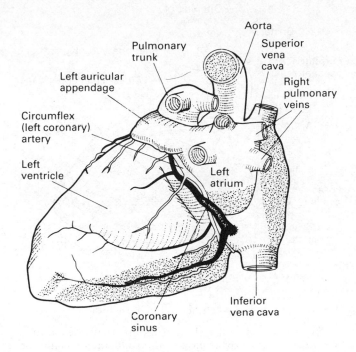

Fig. 40 Posterior surface of heart.

Cavities of the heart (Fig. 41)

Right atrium
Consists of a principal cavity and an appendage, the *auricle*.
The auricle is a small muscular pouch projecting to the left
from the anterior, upper angle of the atrium and overlaps
the ascending aorta and the root of the pulmonary trunk.
Its interior is marked by parallel muscular ridges (*musculi
pectinati*).

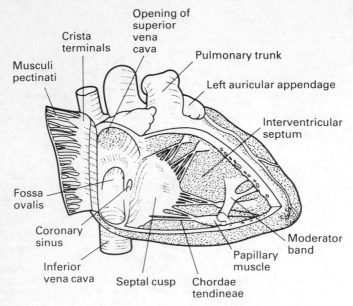

Fig. 41 Interior of right atrium and right ventricle.

The main cavity of the right atrium is divided into a smooth posterior part (*sinus venarum*) and a ridged anterior part by an external vertical groove (*sulcus terminalis*), which on the interior appears as a ridge (*crista terminalis*). The ridge and groove are on the right border of the heart.

Superior vena cava opens into upper and *inferior vena cava* into lower part of the sinus. Between the openings of the two cavae on the posterior wall is a small muscular projection, the *intervenous tubercle*.

The *coronary sinus* opens into the right atrium between the inferior vena cava and the *atrioventricular orifice* which leads to the right ventricle and lies below and to the left in front of the inferior vena cava. It is about 3 cm in diameter.

The *valve of the inferior vena cava (Eustachian valve)* is a semilunar fold of endocardium in front of the anterior margin of the inferior vena cava; it passes upwards and to the left to margin of the *fossa ovalis*. The *valve of the coronary sinus (Thebesian valve)* protects the opening of the coronary sinus.

The *fossa ovalis* is a depression on the interatrial septum, above the opening of the inferior vena cava (Fig. 41). The *limbus of the fossa ovalis* is the upper crescentic elevated margin of the fossa. The *foramen ovale* in the fetus (may be present in the adult) is an opening in the upper part of the floor of the fossa ovalis into the left atrium. In the fetus the blood from the inferior vena cava is directed by its valve and the limbus into the left atrium, thus bypassing the lungs. The blood from the superior vena cava passes through the right atrioventricular opening into the right ventricle and bypasses the lungs by going from the pulmonary artery to the arch of the aorta through the *ductus arteriosus* (p. 142).

Right ventricle
Consists of a cavity, the upper angle of which is prolonged into a funnel-shaped canal, the *infundibulum*, leading to the pulmonary trunk. On the wall (except in the infundibulum, which is smooth) are projections, *trabeculae carneae* consisting of muscular bundles. There are three types. The first are prominent ridges attached to the wall along their whole length. The second are attached at the ends, and are free in the middle. Of these the *septomarginal (moderator) band* extends from the septum to the base of the anterior papillary muscle; it may assist in preventing over-distention of the ventricle, and it conducts the right branch of the atrioventricular bundle to the anterior wall of the ventricle. The third are the conical *papillary muscles* (anterior and posterior) which project inwards and are attached by their bases to the wall of the ventricle. The other end is connected to the *chordae tendineae*, cords attached to the *cusps* of the right atrioventricular valve (Fig. 41).

The *atrioventricular (tricuspid) valve*, which guards the right atrioventricular orifice, consists of three cusps, formed by a reduplication of the endocardium, with some fibrous tissue enclosed. The bases of the flaps are attached to the fibrous ring of the right atrioventricular orifice, and their free edges and ventricular surfaces to the chordae tendineae. The *anterior cusp* is related to the inside of the sternocostal surface of ventricle. The *septal* is posterior and lies on membranous part of interventricular septum. These two are equal in size. Smaller *inferior cusp* lies on diaphragmatic wall of ventricle. The valve prevents regurgitation of blood into the atrium during ventricular contraction.

The *opening of the pulmonary trunk* is at the summit of the funnel-shaped infundibulum, and is guarded by the *pulmonary valve* which consists of three semilunar folds, a right, left and posterior, which guard the orifice of the pulmonary trunk. The free margin of each has in its middle a small fibrous *nodule*. Between each valve and the beginning of the pulmonary trunk is a dilatation called the *pulmonary sinus*.

Left atrium

Consists of a principal cavity, the lining of which is smooth, and an *auricle*. The latter extends forwards and to the right, overlapping the commencement of the pulmonary trunk. Its interior has muscular ridges as on the right side.

The four *pulmonary veins* open into the sides of the cavity, two from each lung, *superior* and *inferior*. The *atrioventricular orifice* is about 2 cm in diameter and lies inferiorly.

Left ventricle

Longer and more conical than the right, with walls nearly three times as thick. The interior contains *trabeculae carneae papillary muscles (superior* and *inferior)*, and *chordae tendineae*, as on the right side. Related to the left atrioventricular orifice is the *left atrioventricular (mitral, bicuspid) valve*, which is attached to the circumference of the opening. It has two flaps, larger *anterior* and smaller *posterior*. The anterior is often called the *aortic* because it is near the opening into the aorta. The *aortic opening* is placed in front and to the right side of the valve. The orifice is guarded by the *aortic valve* which is similar to that of the pulmonary artery with three semilunar folds, *nodules* and *sinuses*. The cusps are named from their position anterior, right posterior and left posterior.

The *endocardium* is the endothelial membrane lining the whole of the interior of the heart, and is continuous with the lining membrane of the blood vessels.

Arteries and veins of the heart (Figs. 38, 39 and 41)

The *coronary arteries* supply the walls of the heart.

Right coronary artery
Comes from the anterior aortic sinus, passes forwards on right side of pulmonary trunk, between it and the right auricle, then downwards, then backwards in the coronary sulcus as far as the inferior interventricular groove. Here it divides into two branches, one of which continues in the coronary groove and anastomoses with the left artery. The other passes to the left as the *posterior interventricular artery* to supply the ventricles and septum and anastomose with the anterior interventricular artery.

Left coronary artery
Arises from the left posterior aortic sinus; passes behind and then to left of pulmonary artery between it and left auricle and divides into two branches, one of which passes forwards and downwards in the anterior interventricular groove (*anterior interventricular artery*) and the other to the left and backwards in the coronary sulcus. This branch, formerly regarded as the continuation of the left coronary artery, is now called the *circumflex artery*.

The right coronary artery gives off a *right marginal branch* which runs along the lower border of the right ventricle. The circumflex artery gives off a *marginal branch* which runs along the left border of the heart (the left ventricle).

Cardiac veins
Return the blood from the muscular wall mainly through the coronary sinus into the right atrium (Fig. 39). The *great cardiac vein* runs in the anterior interventricular groove from apex to base of ventricles, curves to left side and back part of heart and empties into coronary sinus. It is guarded by two valves and receives *left marginal vein*.

The *middle cardiac vein* runs in posterior interventricular groove and terminates in coronary sinus. The *posterior vein of the left ventricle* to left of middle cardiac ends in coronary sinus.

The *small cardiac* runs in the coronary sulcus to right of coronary sinus which it joins. It receives the *right marginal vein*.

The *coronary sinus* is the terminal part of the great cardiac vein which lies in the left part of the coronary sulcus. It is 2–3 cm in length and ends in the right atrium, the opening being guarded by the *valve of the coronary sinus*. It receives tributaries as above, and a small straight vein at the back of the left atrium, the *oblique vein of left atrium*, the remnant of the left superior vena cava of the fetus.

The *anterior cardiac vein* drains sternocostal surfaces of right atrium and ventricles and opens directly into right atrium. May receive small cardiac vein. It and *venae cordis minimae* enter right atrium independently and coronary sinus carries only 60% of venous drainage of myocardium.

The *venae cordis minimae* are very small veins of the heart wall which end in all the chambers. They are most numerous in the walls of the right atrium.

Conducting system

Sinu-atrial node lies in sulcus terminalis at junction of superior vena cava with right atrium and extends to the left in front of the caval opening. Large ill-defined area containing autonomic nerve fibres. *Atrioventricular node* lies in interatrial septum just above opening of coronary sinus; small and discrete. *Atrioventricular bundle (of His)* from A-V node crosses atrioventricular fibrous septum and membranous part of interventricular septum. Branches to right and left along interventricular muscular septum to walls of ventricles. The nodes and bundle consist of modified cardiac muscle. The S-A node is called the *pacemaker of the heart* because it determines the rate at which the heart beats.

The pericardium

Fibrous pericardium
The heart and roots of great arteries and veins are contained in a fibrous bag, conical in shape. Apex attached

around great vessels about 5 cm from heart; base fused with central tendon of diaphragm. It lies in middle mediastinum. Sensory supply from phrenic and intercostal nerves.

Relations of fibrous pericardium
   Anterior: Sternum and left costal cartilages (4 to 7), remains of thymus, anterior edges of lungs and pleurae.

   Posterior: Oesophagus, descending aorta, 5th–8th thoracic vertebrae.

   Lateral: Lung, root of lung, phrenic nerve on each side.

Serous pericardium
Fibrous pericardium is lined by *parietal layer of serous pericardium*; this is reflected to cover heart and roots of great vessels as the *visceral layer of serous pericardium*. At the reflections veins surrounded by one sleeve and arteries by another sleeve of serous pericardium; between the venous and arterial reflections embryonic dorsal mesocardium has broken down to produce *transverse sinus* of pericardium which lies behind the aorta and pulmonary trunk and behind the atria. Oblique sinus of pericardium is behind the heart, bounded on the right by the venae cavae and above by the left atrium which separates transverse and oblique sinuses. Sensory supply of parietal layer by phrenic nerve. Visceral layer supplied by autonomic sensory nerves (cf. peritoneum and pleura).

THE ARTERIES OF THE THORAX

THE PULMONARY TRUNK

The pulmonary trunk conveys de-oxygenated blood from the right side of the heart to the lungs. It is about 5 cm long and at its origin from the right ventricle, is to the left and in front of the aorta in a common tube of serous pericardium. It is directed upwards and then backwards. It reaches the concavity of the aortic arch, where it divides in front of the left bronchus into its right and left branches. Near the bifurcation a fibrous cord, the *ligamentum arteriosum*, passes from the left branch to the inferior surface of the aortic arch. It is the remains of the fetal *ductus arteriosus* which enables the blood to bypass the lungs. Three-quarters of trunk is within fibrous pericardium; terminal 0.5 cm bifurcates outside fibrous bag.

Branches
Right pulmonary artery is longer than the left, passes in front of oesophagus to the hilum of the right lung, behind the ascending aorta and superior vena cava, where it divides into three primary branches, one for each lobe.

Left pulmonary artery is connected at origin with the arch of the aorta by the ligamentum arteriosum, and passes in front of the descending aorta and left bronchus to the hilum of the left lung, where it divides into two primary branches for the two lobes.

THE AORTA (Figs 38, 40 and 42)

Large main trunk of systemic arteries, situated partly in thorax and partly in abdomen; commences at left ventricle, arches over root of left lung, descends in front of vertebral column, behind diaphragm into abdomen and ends opposite left side of body of 4th lumbar vertebra by bifurcating into *common iliac arteries*. Conveniently divided into four parts; *ascending aorta, arch of aorta, descending thoracic aorta, abdominal aorta*.

Ascending aorta

Extent and course
About 5 cm long. Extends from base of left ventricle at level of lower border of 3rd left costal cartilage, upwards and to the right to level of 2nd right costal cartilage. At its root has three bulges, the sinuses of aorta, one opposite each aortic cusp. With pulmonary trunk makes a spiral in common tube of serous pericardium in front of transverse sinus.

Branches
The right and left coronary arteries (p. 139). Openings are at level of upper border of cusp or higher.

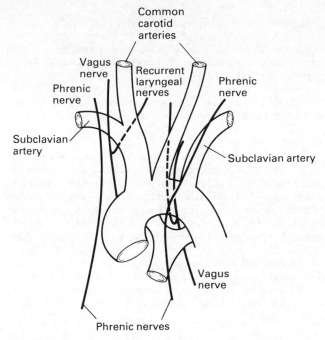

Fig. 42 Relation of vagus and phrenic nerves to arch of aorta and its large branches.

Arch of aorta

Lies wholly in superior mediastinum and passes as much backwards as to the left. Commences at manubriosternal joint (sternal angle) and crosses to left side of 4th thoracic vertebra.

In the concavity of the arch are contained root of left lung, division of pulmonary trunk, ligamentum arteriosum, cardiac nerve plexuses, left recurrent laryngeal nerve, oesophagus and thoracic duct.

Relations

Anterior and to left: Manubrium; thymus; left pleura and lung; left phrenic, cardiac, vagus nerves; left superior intercostal vein.

Superior: Left brachiocephalic vein.

Inferior: Bifurcation of pulmonary trunk; ligamentum arteriosum; left bronchus; left recurrent laryngeal nerve; superficial cardiac plexus.

Posterior and to right: Tracheal bifurcation; deep cardiac plexus; oesophagus; left recurrent laryngeal nerve.

Branches
(1) *Brachiocephalic artery:* About 4 cm long. Ascends to right behind sternum and divides behind right sternoclavicular joint into right common carotid and right subclavian arteries (pp. 146, 157). Lies behind manubrium and origins of right sternohyoid and sternothyroid muscles, thymus gland and left brachiocephalic and right inferior thyroid veins. Trachea at first behind, but to left side above, where artery lies on pleura. On the right are the right brachiocephalic vein and right phrenic nerve; on the left the left common carotid inferiorly and the trachea superiorly. The brachiocephalic artery occasionally gives off a branch, the *arteria thyroidea ima*, which passes in front of trachea to thyroid gland.

(2) *Left common carotid* (p. 146).

(3) *Left subclavian:* Arises behind left common carotid, ascends to neck in contact with apex of left lung, over which it arches. Relations in thorax as those of left common carotid (p. 146).

The descending thoracic aorta

Extends from lower border of 4th thoracic vertebra (left side) to aortic opening in front of body of 12th thoracic vertebra. Lies in posterior mediastinum; is at first to left of bodies of vertebrae, but inferiorly lies in front of them. *Anteriorly* are root of left lung, pericardium and diaphragm; crossed anteriorly by oesophagus at lower end.
Posteriorly are the bodies of vertebrae and hemiazygos veins.

On left are left lung and pleura and *on right* oesophagus above, thoracic duct, and vena azygos; right lung and pleura in lower part but not in actual contact.

Branches

(1) *Bronchial:* Supply the bronchial tree. For the left lung two branches come off from front of aorta. The artery supplying right lung arises either with or from superior left branch, or from first right aortic intercostal.

(2) *Pericardial.*

(3) *Oesophageal* (4 or 5): From front of aorta, running obliquely downwards to supply oesophagus; anastomose with one another, inferior thyroid above and left gastric (below).

(4) *Mediastinal.*

(5) *Posterior intercostal* (9 pairs): Arise from posterior part of aorta, run transversely outwards on bodies of vertebrae, and behind pleura to lower nine intercostal spaces. The right, crossing over front of vertebral column also supply bodies of vertebrae, and pass behind oesophagus, thoracic duct, and azygos veins. The arteries of both sides are crossed anteriorly by sympathetic trunk and its splanchnic branches. On reaching intercostal spaces they divide into anterior and posterior branches; the anterior branch crosses the space obliquely upwards to lower border of the upper rib near the angle; passes forwards in intercostal space in neurovascular plane between internal intercostal and innermost intercostal muscles (p. 122); anastomoses with intercostal of internal thoracic and thoracic branches of axillary. Above the artery is intercostal vein, and below intercostal nerve. A branch, the collateral intercostal, is given off near the angle of the rib, and runs along the upper border of the lower rib. Branches accompany the lateral cutaneous nerves of the thorax from the main trunks of the intercostals. The three lower branches pass forwards between muscles of abdominal wall and anastomose with epigastric and lumbar.

The posterior branch passes backwards between vertebrae and gives off a spinal branch entering intervertebral foramen supplying cord, membranes and body of vertebra, and a muscular branch, which supplies muscles and skin of back.

(6) *Subcostal arteries* (corresponding to intercostal below 12th rib): On each side passes into abdomen behind lateral arcuate ligament and follows lower border of 12th rib.

The 1st and 2nd intercostal arteries come from costo-cervical branch of subclavian (p. 161).

Note: The internal thoracic is a branch of the subclavian artery (p. 160) and the superior and lateral thoracic and acromiothoracic are branches of the axillary (p. 163).

THE ARTERIES OF THE HEAD AND NECK

THE COMMON CAROTID ARTERY

Arises on the right side from the brachiocephalic at its bifurcation behind the right sternoclavicular joint. On the left it arises from the highest part of aortic arch, is longer than the right artery, and is deeply placed in the thorax at its origin. It ascends obliquely to the neck.

Relations
(1) *Left common carotid in thorax:*

Anterior: Left brachiocephalic vein; thymus; overlapping pleura and lung; manubrium.

Medial: Trachea and oesophagus; thoracic duct; left recurrent laryngeal nerve; brachiocephalic artery.

Lateral: Left subclavian artery (also posterior); left vagus and phrenic nerves; left pleura and lung.

(2) *Both common carotids in neck:*

Anterior: Skin; superficial fascia containing platysma; superficial (investing) layer of deep cervical fascia; overlapped inferiorly by sternocleidomastoid; crossed by omohyoid, superior and middle thyroid and anterior jugular veins; ansa cervicalis.

Lateral: Internal jugular vein becoming anterior near sternoclavicular joint.

Posterior: Transverse processes of lower cervical vertebrae (can be pressed on to 6th); prevertebral muscles on each side (scalenus anterior lateral and longus muscles medial); inferior thyroid and vertebral arteries; sympathetic trunk, recurrent laryngeal and vagus nerves.

Medial: Trachea and larynx with pharynx and oesophagus behind them; thyroid gland; superior thyroid artery.

Note: In the neck the common carotid extends from sternoclavicular joint to level of upper border of thyroid cartilage (3rd cervical vertebra) where it divides into external and internal carotids. It is indicated by the lower part of a line drawn from sternoclavicular joint to a point midway between mastoid process and angle of mandible. Carotid sheath encloses internal carotid artery (medial), internal jugular vein (lateral) and vagus nerve (between and behind them).

Branches: Terminal only—*external* and *internal carotids*. At its division the common carotid widens; this dilatation, the *carotid sinus*, extends upwards into the internal carotid artery; its wall is thinner than elsewhere, and has many sensory nerve filaments from the glossopharyngeal, vagus and sympathetic (baroreceptors).

THE EXTERNAL CAROTID ARTERY

Extends from bifurcation of common carotid at level of upper border of thyroid cartilage, to neck of mandible where embedded in the parotid gland it divides into superficial temporal and maxillary. Surrounded by plexus of sympathetic fibres whose cell bodies are in superior cervical ganglion. At first medial then lateral to internal carotid.

Relations:
Superficial: Skin; superficial fascia with platysma; deep cervical fascia; sternocleidomastoid deep to which crossed by stylohyoid and posterior belly of digastric; deep in

parotid gland in which retromandibular vein and facial nerve are superficial to artery; crossed by facial and lingual veins; hypoglossal nerve is superficial to external and internal carotids.

Deep: Styloid process with styloglossus and stylopharyngeus; glossopharyngeal nerve and pharyngeal branch of vagus (between external and internal carotids); superior laryngeal nerve (deep to both external and internal carotids).

Branches

(1) *Ascending pharyngeal:* Small and arises just above bifurcation of common carotid; runs upwards deep to internal carotid and stylopharyngeus alongside pharynx to base of skull. Supplies prevertebral muscles, pharynx, auditory tube and a palatine branch to soft palate and tonsil; also branches to middle ear and meninges.

(2) *Superior thyroid:* Arises just below greater horn of hyoid; curves downwards and forwards to upper pole of thyroid gland accompanied by external laryngeal nerve. Gives off muscular branches including sternocleidomastoid branch which passes downwards and laterally across sheath of common carotid and supplies muscle and skin. Named branches include *infrahyoid* running transversely along inferior border of hyoid and *superior laryngeal* with internal laryngeal nerve, piercing thyrohyoid membrane and supplying muscles, glands and mucous membrane of pharynx and larynx down to vocal folds.

The thyroid branches supply the upper pole and anterior surface within pretracheal fascial sheath; anastomose with opposite artery at isthmus of gland and with inferior thyroid artery at sides of lobes.

(3) *Lingual:* Arises from anterior part of external carotid between superior thyroid and facial arteries; loops upwards and forwards on middle constrictor of pharynx to tip of greater horn of hyoid where it is crossed by hypoglossal nerve; thence forwards deep to hyoglossus above mylohyoid parallel with hypoglossal nerve; finally on genioglossus

upwards and forwards to tip of tongue as *profunda artery of tongue.*

It gives off a *suprahyoid branch*, two or three branches (*dorsal lingual*) ascending to dorsum of tongue and supplying tonsil and soft palate as well as tongue, *sublingual* to supply sublingual gland, adjacent muscles and mucous membrane, and *profunda artery of tongue*, a continuation of the lingual to tip of tongue, accompanying lingual nerve.

(4) *Facial:* Arises near angle of mandible, loops forwards on middle constrictor deep to posterior belly of digastric inferior to mylohyoid; thence into groove on deep surface of submandibular gland to body of mandible; then downwards between gland and mandible; finally around inferior border of mandible, over which it ascends to face, anterior to masseter muscle. Can be felt pulsating on lower border of mandible 4 cm anterior to angle at anterior border of masseter. It ascends tortuously to angle of mouth, then to junction of ala of external nose to cheek, then to medial angle of orbit where it terminates as the angular artery by anastomosing with dorsalis nasi of ophthalmic. It rests successively upon mandible, buccinator, and levator anguli oris and is deep to platysma, risorius, zygomaticus major, and usually levator labii superioris. Facial vein is straight and is posterior to artery.

Branches in the neck: Ascending palatine ascends between styloglossus and stylopharyngeus and pierces pharyngobasilar fascia supplying muscles, tonsil, auditory tube, soft palate, glands, etc., and anastomosing with tonsillar and artery of opposite side. *Tonsillar* penetrates superior constrictor of pharynx to supply tonsil and root of tongue. Branches to submandibular gland. *Submental* arises as artery turns round lower border of mandible, runs forwards over mylohyoid, supplying it and digastric, and divides into superficial branch which supplies lower lip and a deep branch which perforates mylohyoid and supplies floor of mouth and sublingual gland.

Branches in the face: Muscular to masseter, buccinator, etc.; *Inferior labial:* arises near angle of mouth; has a tortuous course between mucous membrane of lower lip and orbicularis oris; anastomoses freely with opposite artery. *Superior labial:* arises with or near preceding, having corresponding course in upper lip; anastomoses with opposite artery and supplies a branch to septum of nose. *Lateral nasal:* supplies ala and side of nose and anastomoses with opposite artery, nasal branch of ophthalmic, and infraorbital branch of maxillary.

Note: Facial and lingual often arise by a common trunk.

(5) *Occipital:* Arises from posterior part of artery, opposite facial, runs upwards and backwards along lower border of posterior belly of digastric deep to mastoid process and its attached muscles in occipital groove; lies on rectus capitis lateralis, obliquus capitis superior and semispinalis capitis; crosses trapezius near attachment to occipital bone and ascends with greater occipital nerve to the back of scalp. The artery crosses internal carotid, accessory nerve, and internal jugular vein; the hypoglossal nerve loops round it near its origin from external carotid.

Branches: Muscular to digastric, stylohyoid and sternoc-leidomastoid (guide to accessory nerve); *mastoid* through mastoid foramen to dura mater; *meningeal* through jugular foramen to dura mater in posterior fossa; *descending branch* to back of neck dividing into superficial branch beneath splenius, supplying it and the trapezius and deep branch going deep to semispinalis capitis, and anastomosing with vertebral, deep cervical of costocervical trunk; *occipital* to muscles and skin of occiput.

(6) *Posterior auricular:* Arises just above occipital artery and runs backwards along upper border of posterior belly of digastric to groove between auricle and mastoid process; supplies auricle and scalp.

Stylomastoid branch is important; enters stylomastoid foramen and supplies facial nerve, middle ear and mastoid antrum and air cells.

(7) *Superficial temporal:* Smaller of two terminal branches; continues in the line of external carotid and is embedded at first in parotid gland; crosses over root of zygoma in front of auriculotemporal nerve where it can be felt pulsating and runs upwards deep to skin for 5 cm; divides into anterior and posterior terminal branches.

Branches: To parotid gland; *transverse facial* arises in parotid, accompanies transverse branches of facial nerve and parotid duct across face; lies above the duct and supplies muscles, glands etc.; supplies temporalis muscle (grooves squamous temporal), auricle, side of scalp.

(8) *Maxillary:* Arises in parotid; mandibular part curves forwards between mandible and sphenomandibular ligament, parallel to auriculotemporal nerve; lies on medial pterygoid muscle and inferior alveolar nerve; pterygoid part runs forwards and laterally on lateral surface of lower head of lateral pterygoid (may go deep to lower head of muscle and is therefore medial to it); pterygopalatine part enters pterygopalatine fossa between two heads of origin of lateral pterygoid.

Branches: Correspond with branches of mandibular and maxillary nerves with some additional branches. *From mandibular part:* branches to middle ear (*anterior tympanic*) and external ear (*deep auricular*).

An important branch is the *middle meningeal* which arises between sphenomandibular ligament and neck of mandible, passes under lateral pterygoid between two roots of auriculotemporal nerve, and enters skull through foramen spinosum of the sphenoid; divides on greater wing of sphenoid into *anterior* and *posterior branches*, anterior going to anterior inferior angle of parietal, posterior to squamous temporal; anterior branch passes almost vertically upwards and posterior branch almost horizontally backwards.

Note: Anterior branch of middle meningeal artery is liable to be torn in fractures of vault of skull as fractures commonly pass through the thin part of the skull where this artery lies. This area is known as *pterion*, 4 cm above middle of zygomatic arch where parietal, frontal, squamous temporal and greater wing of sphenoid meet.

The meningeal arteries supply the diploe of the bones of the vault and give off branches to trigeminal ganglion, middle ear and orbit.

Accessory meningeal passes through foramen ovale, supplies trigeminal ganglion and dura mater. (Generally a branch of the middle meningeal.)

Inferior alveolar descends with and behind inferior alveolar nerve through mandibular canal and supplies teeth and gums; divides opposite first premolar into incisive and mental, the former going to incisor teeth, the latter coming out through mental foramen. Inferior alveolar gives off mylohyoid branch as artery enters foramen; runs in mylohyoid groove.

From pterygoid part: branches to muscles of mastication (*deep temporal* to temporalis muscle and scalp; to pterygoid muscles; to masseter through mandibular notch with nerve to masseter) and to cheek (*buccal* with buccal nerve from mandibular).

From pterygopalatine part: posterior superior alveolar given off as maxillary artery passes into pterygopalatine fossa; descends on posterior aspect of maxilla with branch of maxillary nerve; branches enter posterior dental canals, supplying molars, premolars, maxillary sinus, gums, etc.

Greater palatine passes downwards through greater palatine canal with greater palatine nerve, emerges at posterolateral angle of hard palate and then forwards in a groove lateral to the nerve, and into incisive fossa.

Lesser palatine accompanies greater palatine and then through lesser palatine foramina to muscles and mucosa of soft palate.

Sphenopalatine enters nasal cavity through sphenopalatine foramen and supplies posterior ethmoidal cells, septum, etc.

Infra-orbital is continuation of main trunk; accompanies infra-orbital nerve through infra-orbital canal and appears on face through infra-orbital foramen. In the canal it gives off branches to orbit and *middle* and *anterior superior alveolar* branches to supply front teeth. On the face infra-orbital supplies lacrimal sac and medial angle of orbit as well as lower eyelid and cheek.

THE INTERNAL CAROTID ARTERY

Extends from bifurcation of common carotid at level of upper border of thyroid cartilage to base of brain where it divides into its terminal branches, the anterior and middle cerebral arteries. Accompanied by postganglionic sympathetic plexus of nerve fibres (*carotid nerve*) whose cell bodies are in the superior cervical ganglion. In its course the artery passes directly upwards from common carotid to carotid canal of temporal bone; in canal it passes upwards, medial to and then above auditory tube, then forwards and medially, then forwards into cavernous venous sinus, with the abducent nerve on its lateral side; it turns upwards medial to anterior clinoid process, pierces dura mater, passes backwards on roof of cavernous sinus and divides between optic and oculomotor nerves lateral to optic chiasma below anterior perforated substance into terminal branches.

Relations in the neck
Superficial (anterior): Skin, superficial fascia with platysma; deep cervical fascia; overlapped by sternocleidomastoid; deep to parotid gland where it reaches the styloid process and its attached muscles—styloglossus, stylopharyngeus, stylohyoid; last crosses artery with posterior belly of digastric; occipital and posterior auricular arteries cross the beginning of the internal carotid where hypoglossal nerve crosses the artery; glossopharyngeal and pharyngeal branch of vagus pass forwards between internal and external carotids.

Posterior: Transverse processes of cervical vertebrae (4th to 1st) and longus capitis; sympathetic trunk, vagus (posterolateral); internal jugular vein posterior at base of skull, becomes lateral lower down.

Medial: Pharynx; ascending pharyngeal artery; superior laryngeal nerve.

Lateral: Internal jugular vein.

Branches
From petrous part: caroticotympanic through foramina in carotid canal to middle ear; *From cavernous part: hypophyseal* to pituitary gland; *meningeal* to dura mater in middle cranial fossa. The *ophthalmic artery* is a large and important branch.

It arises after internal carotid has pierced the dura mater at medial side of anterior clinoid process and enters orbit through optic canal lateral to and below optic nerve; it then crosses superior to nerve to its medial side and runs forwards between superior oblique and medial rectus to the front of the orbit and divides into its terminal branches, supratrochlear and dorsalis nasi.

Branches of ophthalmic artery: Lacrimal accompanying lacrimal nerve over lateral rectus to lacrimal gland; gives off zygomatic branches and a branch which enters cranial cavity through superior orbital fissure.

Central artery of retina comes off before ophthalmic artery becomes lateral to optic nerve and continues forwards inferior to optic nerve into which it sinks 1 cm behind eyeball; runs in its substance to retina (p. 322); an end-artery.

Posterior ciliary: Two sets—*short* (perforate sclera and supply choroid) and *long* (2) (pass forwards between choroid and sclera).

Supra-orbital: Ascends and accompanies frontal nerve on levator palpebrae superioris to pass through supra-orbital notch; ascends over frontal bone and supplies scalp.

Muscular, superior and *inferior:* To muscles of orbit, give off anterior ciliary (6–8), which pierce sclera behind cornea.

Ethmoidal: Anterior and *posterior* to ethmoidal cells, through anterior and posterior ethmoidal foramina respectively, supplying also dura mater; the anterior accompanies external nasal branch of anterior ethmoidal nerve (p. 253) to skin of nose (anterior nasal artery).

Medial palpebral (2): One for each lid, arise near pulley for superior oblique, form an arch in each lid, and supply lacrimal apparatus.

Supratrochlear: Turns upwards round medial margin of orbit to scalp.

Dorsalis nasi: Over medial palpebral ligament to root of nose.

From cerebral part: Anterior cerebral: arises at medial extremity of lateral cerebral sulcus, passes forwards and medially across anterior perforated substance, above optic nerve; here united with opposite artery by anterior communicating; passes round genu of corpus callosum into longitudinal fissure; passes backwards above corpus callosum. *Branches: central* to anterior perforated substance; *cortical* to orbital surface of frontal lobe, medial surface of hemisphere as far back as parieto-occipital sulcus, overlapping superomedial border for 2 cm.

Middle cerebral: Largest branch, enters lateral cerebral sulcus and passes to lateral surface. *Branches: central* to thalamus and corpus striatum, one of which is called *artery of cerebral haemorrhage*; *cortical* to insula, opercula and lateral surfaces of frontal, parietal and temporal lobes to within 2 cm of their borders.

Posterior communicating: From posterior part of internal carotid, runs backwards to anastomose with posterior cerebral of basilar to complete arterial circle.

Anterior choroidal: From deep aspect of artery, passes backwards over cerebral peduncle and enters inferior horn of lateral ventricle; supplies hippocampus, fimbria, and choroid plexus.

THE BASILAR ARTERY

Formed by union of the two vertebral arteries (p. 158) each of which is a branch of subclavian artery. Vertebral arteries enter skull through foramen magnum. Basilar extends from lower to upper border of pons, where it divides into two posterior cerebral arteries.

Branches
Pontine to pons; *labyrinthine* to internal ear, with vestibulocochlear nerve; *anterior inferior cerebellar* to anterior part of

inferior surface of cerebellum; *superior cerebellar* arises near termination, passes laterally below oculomotor nerve to wind round cerebral peduncle with trochlear nerve and basal vein (p. 185), and supplies upper surface of cerebellum; *posterior cerebral*, one on each side, terminal of basilar, winds backwards round cerebral peduncle above oculomotor nerve, passes upwards to under surface of occipital lobe, gives off *central* to inside of hemisphere (especially to thalamus and internal capsule), *choroidal* to choroid plexus through choroid fissure, *cortical* (especially to visual cortex on medial surface of occipital lobe). Supplies whole of occipital lobe, inferior surface of temporal lobe except operculum.

THE CEREBRAL ARTERIAL CIRCLE (OF WILLIS)

This refers to the anastomoses between the *vertebral* and *internal carotid arteries* at base of brain in the region of the

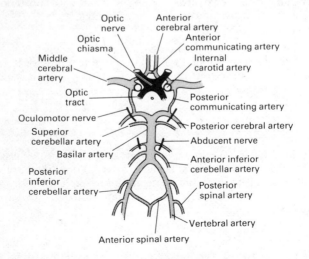

Fig. 43 Cerebral arterial circle (of Willis).

optic chiasma and pituitary stalk. The *anterior cerebral arteries* are connected by the *anterior communicating*. Each *posterior cerebral artery* is joined to an internal carotid by a *posterior communicating artery* (Fig. 43).

THE SUBCLAVIAN ARTERIES

The right arises from the brachiocephalic artery at level of sternoclavicular joint and the left from convexity of arch of aorta (p. 144). Each passes into neck, arches laterally over pleura, lies on 1st rib between scalenus anterior and medius, and ends at lateral border of rib.

The *scalenus anterior*, passing anterior to artery, divides it into three parts—1st part from origin to medial border of scalenus anterior, 2nd part posterior to scalenus anterior, 3rd part from lateral edge of muscle to lateral border of 1st rib.

Relations

Subclavian vein is anterior and inferior to artery but lies in front of scalenus anterior. The lower ends of *sternocleidomastoid*, *sternohyoid* and *sternothyroid muscles* are anterior to first part. *Internal jugular vein* passes anterior to first part as it joins the subclavian vein to form the brachiocephalic vein. The *vertebral* (vertical) and *anterior jugular* (horizontal) veins are also anterior to the first part. The *pleura and lung* are posterior to the first and second parts of the artery and the *trunks of the brachial plexus* and *scalenus medius* are posterior and superior to the third part.

The *phrenic nerve* runs downwards on scalenus anterior and on the right crosses second part before entering thorax. The *vagus nerve* on right is anterior to first part and gives off recurrent laryngeal nerve which hooks below subclavian artery and passes upwards behind it. On left, vagus lies between common carotid and subclavian. Both left phrenic and left vagus in thorax are in front of subclavian artery. *Inferior cervical ganglion* and *sympathetic trunk* are posterior to the first part and *ansa subclavia* from middle to inferior cervical ganglion is anterior to and loops below and then behind the first part. *Cardiac branches* of vagus and sympathetic pass downwards into the thorax in relation to first part.

Since the *left* subclavian artery arises from the aorta, the *first part in the thorax* is related to the left common carotid, trachea, oesophagus, with left recurrent laryngeal nerve between them, thoracic duct and vagus on its medial side. The *thoracic duct* arches forwards and laterally to cross anterior to the first part in the neck and enter the beginning of the left brachiocephalic vein.

Anterior to the third part there are the *suprascapular, transverse cervical, external jugular* and *anterior jugular veins* on their way to the subclavian vein (the *suprascapular nerve* and *artery* accompany the vein) the *subclavius muscle* and its nerve and the *clavicle*.

Note: Third part of the artery lies in the greater supraclavicular triangle ensheathed in fascia; it carries a sheath of fascia into the axilla (armpit) as it continues as the axillary artery.

Branches from 1st part of subclavian

Vertebral: First passes upwards and backwards, behind inferior thyroid artery and internal jugular and vertebral veins, then between scalenus anterior and longus cervicis, to enter foramen in transverse process of 6th cervical vertebra. *Secondly*, ascends in corresponding foramina including that of axis with vertebral vein in front and cervical nerves behind. It then passes laterally and upwards, through foramen transversarium of atlas. *Thirdly*, winds backwards and medially behind lateral mass on to groove on posterior arch of atlas, lying in floor of suboccipital triangle, with 1st cervical nerve between it and the bone, and the dorsal ramus of same nerve emerging posteriorly. *Fourthly*, passes deep to posterior atlanto-occipital membrane and pierces dura mater to enter skull through foramen magnum. It then winds round medulla, above ligamentum denticulatum and passes between hypoglossal and ventral root of 1st cervical nerve to front of medulla. Unites with fellow to form the basilar artery (p. 155) at the lower border of the pons. Carries sympathetic fibres from cell bodies in inferior cervical ganglion.

Branches of vertebral in neck
Spinal entering vertebral canal through an intervertebral foramen, supplying posterior root ganglion and nerve roots; *muscular* to deep cervical muscles.

Branches of vertebral in cranium
Posterior meningeal arises before vertebral artery pierces dura mater at foramen magnum, to falx cerebelli and dura in posterior fossa.

Posterior spinal arises near posterior part of medulla, passing downwards on back of cord behind dorsal nerve roots. Supplies posterior grey and posterior white columns. *Anterior spinal* arises near end of artery, descends in front of medulla, unites with fellow just below foramen magnum to form a single artery, which continues along the cord, anastomosing with the posterior artery over conus medullaris; supplies pia mater and lateral and anterior white columns and anterior grey column.

Note: Reinforcement by spinal branches is negligible except at T. 1 and T. 11 levels (*the arteries of Adamkiewicz*) for cervical and lumbar enlargements of cord.

Posterior inferior cerebellar arises near pons with tortuous course between roots of hypoglossal and then between accessory and vagus nerves to reach inferior surface of cerebellum. Supplies hemisphere, vermis, choroid plexus of 4th ventricle and posterior part of medulla.

Thyrocervical trunk
A short thick trunk from front of artery near medial border of scalenus anterior, dividing into:

(1) *Inferior thyroid:* Passes upwards in front of vertebral artery and deep to internal jugular vein; then medially and downwards behind carotid sheath and sympathetic (middle cervical ganglion); finally turns medially to lower part and posterior surface of thyroid gland. Carries sympathetic fibres from cell bodies in middle cervical ganglion.

The inferior thyroid artery gives off (a) *ascending cervical* which arises as inferior thyroid turns behind carotid sheath, ascends medial to phrenic nerve and between scalenus anterior and longus capitis and supplies them, (b) the *inferior laryngeal* which runs upwards on trachea with recurrent laryngeal nerve, (c) *tracheal* and *oesophageal branches*, (d) *glandular branches* which divide outside pretracheal fascia (4 or 5 branches) and pierce it separately (cf. superior thyroid artery, p. 148) (recurrent laryngeal nerve lies close behind these branches).

(2) *Suprascapular:* Runs laterally in front of scalenus anterior and phrenic nerve, just above 3rd part of subclavian; then backwards behind and parallel to clavicle under cover of trapezius to upper edge of scapula; then inclines downwards with nerve and passes over suprascapular ligament on suprascapular notch to enter supraspinous fossa; in contact with the bone deep to supraspinatus, which it supplies; then winds round neck of scapula to infraspinous fossa; anastomoses with circumflex scapular, deep branch of transverse cervical and subscapular arteries, an important anastomosis if the 3rd part of the subclavian is tied; suprascapular artery supplies several muscles including sternocleidomastoid, the shoulder joint, the clavicle and scapula.

(3) *Transverse cervical:* Passes laterally and backwards over scalene muscles, phrenic nerve and brachial plexus, deep to omohyoid, to lateral edge of levator scapulae, where it divides into a superficial branch which ascends deep to trapezius towards occiput, and a deep branch which runs deep to levator scapulae then downwards along medial border of scapula.

Note: Either or both of suprascapular and transverse cervical arteries may arise from third part of subclavian and run between trunks of brachial plexus.

Internal thoracic (mammary) arises from inferior surface of subclavian just below thyrocervical trunk, runs downwards behind clavicle and subclavian vein to posterior surface of 1st costal cartilage, where it is crossed by phrenic nerve; then downwards between pleura and costal cartilages, crossed anteriorly by intercostal nerves and lies posterior to transversus thoracis as far as the 6th intercostal space where it divides into two terminal branches.

The internal thoracic gives off:

(a) *Pericardiacophrenic* which arises high in chest and accompanies phrenic nerve between pleura and pericardium to diaphragm.

(b) *Anterior mediastinal.*

(c) *Anterior intercostal* to upper five or six intercostal spaces, two in each space anastomosing with aortic intercostal arteries; note this important anastomosis between branches from the subclavian, proximal to site of coarctation (narrowing) of aorta, and the aortic intercostal arteries, distal to the narrowing.

(d) *Perforating* which perforate upper five or six intercostal spaces to supply pectoral muscles and mammary gland; branches to breast especially large in second and third spaces.

(e) *Musculophrenic*, lateral of two terminal branches, perforates diaphragm about 9th intercostal space and supplies diaphragm and anterior branches to lower intercostal spaces.

(f) *Superior epigastric*, medial terminal branch, passes behind 7th costal cartilage and pierces diaphragm; lies posterior to rectus muscle within its sheath and terminates in that muscle to anastomose with inferior epigastric of external iliac.

Branches from 2nd part of subclavian
Costocervical trunk: from upper and back part behind scalenus anterior, bends backwards over pleural dome to front of neck of 1st rib, where ventral ramus of 1st thoracic nerve is lateral, and 1st thoracic ganglion of sympathetic is medial to artery. It divides into (a) *superior intercostal* which supplies 1st and 2nd intercostal spaces, small branches to cord and deep muscles of back, (b) *deep cervical* which

passes between transverse process of 7th cervical vertebra and 1st rib and ascends deep to semispinalis capitis to axis.

THE ARTERIES OF THE UPPER LIMB

THE AXILLARY ARTERY

Extends from lateral border of 1st rib to lower border of teres major. Divided into three parts by pectoralis minor.

1st part extends from lateral border of 1st rib to upper border of pectoralis minor.

Relations
    Anterior: Pectoralis major, clavipectoral fascia, cephalic vein, lateral pectoral nerve, and branches of acromiothoracic artery.

   Posterior: 1st intercostal space and muscles, 1st serration of serratus anterior, nerve to serratus anterior, medial cord of brachial plexus.

   Medial: Medial pectoral nerve and axillary vein.

   Lateral: Posterior and lateral cords of brachial plexus.

2nd part extends from superior to inferior border of pectoralis minor.

Relations
    Anterior: Pectoralis minor and major.

   Posterior: Subscapularis, posterior cord of plexus.

   Medial: Medial cord of plexus, separating artery from vein.

   Lateral: Lateral cord of plexus.

3rd part extends from inferior border of pectoralis minor to lower border of teres major.

Relations

Anterior: Pectoralis major, medial head of median and medial cutaneous nerve of forearm; inferiorly skin and fasciae.

Posterior: Subscapularis, tendons of latissimus dorsi and teres major, radial and axillary nerves.

Medial: Ulnar nerve, axillary vein, and medial cutaneous nerve of arm.

Lateral: Coracobrachialis, median and musculocutaneous nerves.

Branches

Superior thoracic from 1st part arising at level of 1st intercostal space and supplying pectoral muscles and chest wall.

Acromiothoracic from 1st part and giving branches to deltoid muscle, region of acromion, pectoral muscles and clavicle.

Lateral thoracic from 2nd part and supplying chest wall (pectoral muscles, serratus anterior); in female lateral mammary to mammary gland.

Subscapular from 3rd part and passing downwards on posterior wall of axilla with thoracodorsal nerve to lower angle of scapula. It gives off the circumflex scapular branch which passes backwards through triangular space to infraspinous fossa, and also gives branches to subscapular fossa. Takes part in scapular anastomosis.

Note: Very important anastomoses in the scapular region between branches of the 1st part of the subclavian (p. 160) and 3rd part of the axillary.

Posterior circumflex humeral from 3rd part passing backwards and medial to surgical neck of humerus through quadrilateral space with axillary nerve and supplying deltoid, head of humerus, shoulder joint, teres minor, and long head of triceps; anastomoses with acromiothoracic, anterior circumflex humeral and profunda brachii.

Anterior circumflex humeral from 3rd part passing laterally round surgical neck of humerus.

THE BRACHIAL ARTERY

Extends from lower border of teres major to 1 cm distal to bend of elbow (level of neck of radius); runs along medial border of coracobrachialis and biceps brachii accompanied by two veins. Note that upper half of artery is on medial side of upper arm and lower half on front of elbow.

Relations

Anterior: Skin, fascia, upper part of coracobrachialis, biceps brachii and bicipital aponeurosis, median basilic vein. Crossed anteriorly by median nerve from lateral to medial at insertion of coracobrachialis.

Posterior: Long and medial heads of triceps brachii, lower part of coracobrachialis, brachialis, radial nerve, profunda brachii vessels (near axilla).

Medial: Medial cutaneous nerve of forearm to about middle of arm, ulnar nerve to insertion of coracobrachialis, median nerve from insertion of coracobrachialis to elbow.

Lateral: Coracobrachialis and biceps brachii, median nerve from origin of artery to insertion of coracobrachialis.

Branches

Profunda brachii arises at level of lower border of teres major; winds backwards and laterally with radial nerve in the spiral groove; gives off anterior branch piercing lateral intermuscular septum; ends as posterior branch behind lateral epicondyle; in region of elbow anastomoses with interosseous recurrent and supratrochlear. Supplies triceps brachii, anconeus.

Nutrient to humerus.

Superior ulnar collateral accompanies ulnar nerve, pierces medial intermuscular septum; ends in anastomosis round elbow joint.

Inferior ulnar collateral (supratrochlear) arises 5 cm above elbow joint; courses to hollow between olecranon and medial epicondyle of humerus; anastomoses with arteries round elbow joint.

Muscular to coracobrachialis, biceps, brachialis.

THE RADIAL ARTERY

Extends from bifurcation of the brachial to the deep palmar arch of hand; accompanied by two veins.

Relations in the forearm
  Anterior: Skin, fascia, brachoradialis.

  Posterior: Tendon of biceps brachii, supinator, pronator teres, radial head of flexor digitorum superficialis, flexor pollicis longus, pronator quadratus, lower end of radius.

  Medial: Pronator teres, flexor carpi radialis.

  Lateral: Brachioradialis tendon, and for middle third radial nerve.
  The artery passes along medial border of brachioradialis tendon to carpus; winds laterally and posteriorly round carpus deep to long abductor and short extensor of thumb and radial nerve, on lateral ligament of wrist joint, scaphoid and trapezium; enters palm of hand between the heads of the 1st dorsal interosseous muscle and adductor pollicis and forms deep palmar arch (Fig. 31).

Branches
Radial recurrent ascends between brachialis and brachioradialis, supplying them and the elbow joint.
  Muscular to muscles of radial side of forearm.
  Superficial palmar: Arises when the artery is about to wind round carpus, passes between muscles of ball of thumb and completes superficial palmar arch.
  Anterior carpal arises near lower border of pronator quadratus and passes medially under tendons to anastomose with anterior carpal of ulnar.
  Dorsal carpal arises beneath posterior tendons of thumb and anastomoses with posterior carpal of ulnar, forming *dorsal carpal arch*, which gives off 2nd, 3rd and 4th *dorsal metacarpal* to 2nd, 3rd and 4th metacarpal spaces; each anastomoses with proximal perforating artery of deep palmar arch, and at distal end of metacarpal space gives off distal perforating to join palmar digital artery (Fig. 44).

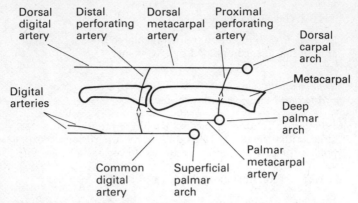

Fig. 44   Arteries of hand.

1st dorsal metacarpal arises near dorsal radial carpal and supplies adjoining sides of thumb and index finger.

Princeps pollicis arises as the artery enters palm, courses between 1st metacarpal and oblique head of adductor pollicis to base of proximal phalanx, where it divides into two terminal branches which run along the sides of the palmar surface of thumb.

Radialis indicis arises near the preceding, passes between 1st dorsal interosseus and transverse head of adductor pollicis to lateral side of index finger; sends a branch to superficial palmar arch.

Deep palmar arch extends from proximal end of 1st metacarpal space to base of 5th metacarpal. It lies over the bases of metacarpal bones and terminates by anastomosing with deep branch of ulnar. The deep palmar arch gives off a *recurrent branch* to the front of the carpus, three *palmar metacarpal branches* which run in the three medial metacarpal spaces and join the digital arteries of the superficial palmar arch at digital clefts, and *proximal perforating branches* (Fig. 44).

THE ULNAR ARTERY

Extends from bifurcation of brachial to superficial palmar arch and passes along lateral side of flexor carpi ulnaris to the palm accompanied by two veins.

Relations in the forearm
Anterior: Pronator teres (both heads), flexor carpi radialis, palmaris longus, flexor digitorum superficialis, proximally median nerve, distally overlapped by flexor carpi ulnaris tendon.

Posterior: Brachialis, flexor digitorum profundus.

Medial: Flexor carpi ulnaris, ulnar nerve.

Lateral: Flexor digitorum superficialis. At wrist lies on flexor retinaculum lateral to the ulnar nerve and pisiform bone.

Branches
Anterior ulnar recurrent arises near beginning of artery, ascends between brachialis and pronator teres, and supplies them.

Posterior ulnar recurrent arises distal to anterior, passes beneath flexor digitorum superficialis, ascends behind medial epicondyle, thence between heads of flexor carpi ulnaris and supplies elbow joint and neighbouring muscles.

Common interosseous about 1 cm long, arises just distal to radial tuberosity, passes towards interosseous membrane and divides into two terminal branches.

Anterior interosseous passes down forearm on anterior surface of interosseous membrane, accompanied by and medial to anterior interosseous branch of median nerve. At upper border of pronator quadratus one branch goes downwards deep to quadratus to anastomose with anterior carpal and deep arch; the other, piercing interosseous membrane, descends to back of carpus and anastomoses with posterior interosseous, posterior carpal of radial and ulnar. Supplies nutrient branches to radius and ulna, muscular branches, and branch to median nerve.

Posterior interosseous passes backwards between oblique cord and interosseous membrane, and between supinator and abductor pollicis longus, runs down back of forearm, lying medial to posterior interosseous nerve, between superficial and deep muscular layers as far as the perforating branch of anterior interosseous; anastomoses with posterior carpal of radial and ulnar, anterior interosseous.

Interosseous recurrent is given off near origin of common interosseous, passes deep to anconeus to interval between olecranon and lateral epicondyle.

Muscular to muscles on ulnar side of forearm.

Anterior and posterior carpal to front and back of carpus. Posterior carpal completes the dorsal carpal arch on medial side.

Superficial palmar arch is continuation of ulnar in the hand, and lies immediately deep to palmar aponeurosis on digital nerves and flexor tendons. It turns laterally just distal to flexor retinaculum, and, forming an arch with the convexity distally, is directed towards the thumb, where the arch becomes completed by superficial palmar of radial or radialis indicis or princeps pollicis. From convex side of the arch it gives off four *palmar digital branches* to supply three medial digits and medial side of index finger.

Deep branch of ulnar artery is given off at commencement of arch, passes downwards with deep branch of ulnar nerve between abductor and short flexor of little finger to complete deep palmar arch (p. 166).

THE ARTERIES OF THE ABDOMEN AND PELVIS

THE ABDOMINAL AORTA

Extends from 12th thoracic vertebra to the left side of front of body of 4th lumbar vertebra where it divides into common iliacs. Enters abdomen between crura of diaphragm in midline but near its bifurcation inclines to left side.

Relations

Anterior (from above downwards): Lesser omentum, stomach, coeliac plexus, pancreas, splenic vein, left renal vein, inferior (3rd) part of duodenum, root of mesentery, coils of small intestine, aortic plexus, peritoneum. It is related to the pancreas and duodenum without any intervening peritoneum.

Posterior: Bodies of upper four lumbar vertebrae, left lumbar veins, cisterna chyli, thoracic duct.

On right: Inferior vena cava, beginning of thoracic duct and vena azygos, right coeliac ganglion, right trunk of sympathetic.

On left: Left coeliac ganglion, left trunk of sympathetic.

Branches (from above downwards):

(1) *Phrenic* (paired): Arise close together immediately inferior to diaphragm, pass on its under surface across crura, the left behind oesophagus, the right behind inferior vena cava. Supplies diaphragm and gives off *superior suprarenal* to suprarenal gland.

(2) *Coeliac trunk* (unpaired): Arises between crura of diaphragm, just above pancreas, 1 cm long, surrounded by coeliac plexus; divides into three visceral branches for supply of foregut.

(i) *Left gastric:* smallest of three branches, directed upwards and to left behind omental bursa (lesser sac of peritoneum) to cardiac end of stomach; gives off a few oesophageal branches, then turns to right along lesser curvature, giving branches to both surfaces of viscus; finally anastomoses with right gastric from common hepatic.

(ii) *Hepatic:* directed to right, forwards and upwards between layers of lesser omentum, and anterior to epiploic foramen (opening into lesser sac) to porta hepatis, lying to left of bile duct and in front of portal vein; at porta hepatis divides into right and left hepatic, supplying corresponding lobes of the liver.

The hepatic gives off: (a) *right gastric* which runs along lesser curvature of stomach from right to left and anastomoses with left gastric; (b) *gastroduodenal* which passes behind superior (1st) part of duodenum and divides into *right gastro-epiploic* which runs along greater curvature of stomach from right to left to anastomose with left gastro-epiploic of splenic and gives off branches to stomach and to greater omentum, and *superior pancreaticoduodenal* which runs downwards between the descending (2nd) part of duodenum and the pancreas to anastomose with inferior pancreaticoduodenal of superior mesenteric; (c) *right hepatic* to right lobe of liver, giving off the *cystic artery* to gall bladder; (d) *left hepatic* to left lobe and also to caudate lobe.

(iii) *Splenic:* directed horizontally on tortuous course to left along upper border of pancreas; crosses the left kidney; reaches spleen by passing between two layers of lienorenal ligament, and gives off gastric branches which reach stomach between layers of gastrosplenic ligament; supplies spleen, stomach and pancreas; divides near spleen into several terminal branches which enter hilum of that viscus. It gives off: (a) *pancreatic branches* as artery runs along pancreas; one of them (*arteria pancreatica magna*) accompanies the duct, running from right to left; (b) *left gastro-epiploic* running in gastrosplenic ligament to right between layers of greater omentum, along greater curvature of stomach and anastomosing with right gastro-epiploic of hepatic; (c) *short gastric branches* (5 or 6) to fundus of stomach in gastrosplenic ligament.

(3) *Middle suprarenal* (paired): Arise a little below coeliac artery; on each side runs transversely outwards over crus of diaphragm to suprarenal gland.

(4) *Superior mesenteric* (unpaired): Artery of midgut (from duodenal papilla to just short of left colic flexure). Arises 1 cm distal to coeliac trunk behind body of pancreas near its neck; lies behind splenic vein and in front of left renal vein. It then passes in front of horizontal (3rd) part of duodenum and curves downwards and to right in front of superior vena cava, right ureter and psoas between the two layers of its mesentery. It terminates in branches to small intestine, caecum and colon. It is surrounded by mesenteric plexus of nerves and accompanied by its vein, which lies to right.

It gives off *inferior pancreaticoduodenal* running from left to right along upper border of horizontal (3rd) part of duodenum.

Middle colic passing forwards in transverse mesocolon from upper part of right side of artery, supplies transverse colon and anastomoses to right with right colic and to left with upper left colic of inferior mesenteric.

Jejunal and *ileal branches* supply jejunum and ileum, twelve or fifteen in number from left convex side of artery; about 5 cm from origin they bifurcate, each division uniting with a neighbouring branch to form an arcade, from which branches issue; these divide and communicate in the same way four or five times, the final branches proceed directly to intestine (more arcades distally, i.e. to ileum).

Right colic from right side of artery to middle of ascending colon, ascending branch anastomoses with middle colic, descending branch with ileocolic.

Ileocolic from right side of artery down to caecum and appendix. Divides into *anterior* and *posterior caecal*; from latter comes *appendicular branch*. Ileocolic gives off a descending branch to lower part of ileum anastomosing with terminal branch of superior mesenteric and an ascending branch to ascending colon to anastomose with right colic.

(5) *Renal* (paired): One from each side arising 1 cm below superior mesenteric, the right a little lower than the left one. Pass laterally to supply kidneys, the right one passing behind inferior vena cava. Each divides near viscus into three branches, which enter hilum two in front and one behind ureter, the last high up. Renal vein lies anterior. It is accompanied by plexus of nerves; supplies branches to suprarenal gland (*inferior suprarenal*), ureter and fatty capsule of kidney.

(6) *Testicular* (paired): Two small, long arteries; each arises just below renal, directed downwards and laterally behind peritoneum over psoas, crossing anterior to ureter, to which it gives branches, and external iliac artery (the right one also crosses the inferior vena cava) to deep inguinal ring; thence accompanied by testicular vein, pampiniform plexus, and ductus (vas) deferens, it passes along inguinal canal, and out of the superficial ring to the scrotum where it divides into branches which enter the posterior surface of the testis; anastomoses (often very poor) with artery of ductus deferens and cremasteric.

In the female the artery is termed *ovarian*, and runs from posterior abdominal wall in suspensory ligament of ovary

with veins, lymphatic vessels and sympathetic plexus between layers of broad ligament of uterus, to ovary and tube;
anastomoses with uterine.

(7) *Inferior mesenteric* (unpaired): Artery of hindgut, from
splenic flexure to mucous membrane at anocutaneous
junction. Arises on left side of aorta, about 3 cm above
bifurcation. Lies at first on left side of aorta, then crosses
over left psoas, left common iliac vessels and ureter, to back
of rectum; supplies left half of transverse colon, descending
and pelvic parts of colon, rectum and part of anal canal.

It gives off *superior left colic* which passes upwards in front
of left kidney supplying descending colon to anastomose
with inferior left colic.

Inferior left colic (sigmoid) to descending and pelvic colon;
anastomosing with superior left colic and superior rectal.

Superior rectal is continuation of inferior mesenteric, passes into pelvis between layers of pelvic mesocolon (giving
branches to lower part of pelvic colon) and divides into two
branches, which pass down, behind rectum, to about 15 cm
from anus, where they subdivide to supply rectum and
anal canal; anastomoses inferiorly with middle and inferior
rectal arteries. At upper limit of its distribution there may
be a gap in arterial anastomotic loops, which is of some
surgical importance.

Marginal artery: anastomoses between ileocolic, right,
middle and left colics result in a continuous arterial line at
margin of gut from caecum to rectum, reinforced by above-
named arteries. Occasionally deficient in places, the marginal artery is usually continuous.

(8) *Lumbar* (4 pairs): From back of aorta, pass laterally,
resting on body of corresponding vertebra deep to sympathetic trunk and psoas; the two upper pairs deep to crura
of diaphragm; the right also behind vena cava. Divide near
transverse processes to supply muscles of abdominal wall.
They pass posterior to quadratus lumborum. A posterior
branch accompanies dorsal ramus of nerve; gives off spinal
branch to supply meninges and spinal nerve roots.

(9) *Median sacral* (unpaired): A small branch given off just

at bifurcation of aorta. It passes downwards over 5th lumbar vertebra and middle of sacrum to coccyx behind left common iliac vein. It is, both morphologically and embryologically, the continuation of abdominal aorta to the tail.

THE COMMON ILIAC ARTERY

Extends from bifurcation of aorta, on body of 4th lumbar vertebra, to brim of pelvis at sacro-iliac joint where it divides into external and internal iliac. About 5 cm long.

Relations
   Anterior: Peritoneum, small intestines, ureter, branches of sympathetic.

   Differences between right and left artery: The right is longer, the aorta being on the left side of vertebra; on right side are inferior vena cava and right psoas. Companion vein at first behind, but to the right at upper part; left common iliac vein crosses to right behind right artery. The left common iliac artery is crossed anteriorly by inferior mesenteric artery, the left common iliac vein is inferior and medial to the artery.

THE INTERNAL ILIAC ARTERY

Extends from bifurcation of common iliac to greater sciatic notch where it divides into anterior and posterior trunks. It is about 3 cm long.

Relations
   Anterior: Peritoneum; ureter runs downwards in front of division of artery, ileum on right, pelvic colon on left.

   Posterior: Medial border of psoas, internal iliac vein, lumbosacral trunk, obturator nerve and sacrum. Note that this artery lies internal to parietal layer of pelvic fascia which is pierced by all the parietal branches e.g. gluteal arteries.
   In fetus internal iliac artery continues as umbilical artery

passing forwards to the side of the bladder and upwards on the abdominal wall to the umbilicus into the umbilical cord to the placenta. After birth the artery becomes obliterated beyond its proximal 3 cm and forms a fibrous cord called the *obliterated umbilical artery* or *lateral umbilical ligament*.

Branches from anterior trunk

Superior vesical arises from unobliterated part of umbilical artery and extends from greater sciatic notch to side of bladder; supplies side and upper part of bladder, ureter, gives off *artery to ductus deferens*.

Inferior vesical to base of bladder (also side of prostate and seminal vesicles).

Middle rectal to prostate, muscle of rectum and, in the female, vagina; may come off with inferior vesical; anastomoses with superior rectal of inferior mesenteric and inferior rectal of internal pudendal (very poor anastomosis).

Uterine passes downwards towards cervix of uterus, then medially between layers of broad ligament, where it crosses the ureter superiorly, and then divides into a large ascending branch to uterus and ovary, anastomosing with ovarian, and small descending branch to vagina.

Vaginal to vagina, base of bladder and lower part of rectum.

Obturator passes downwards and forwards with obturator nerve 1 cm below brim of pelvis to a groove in upper part of obturator foramen; leaves pelvis and divides into two branches at upper border of obturator externus. In pelvis it is placed between pelvic fascia and peritoneum just below obturator nerve, and is crossed medially by ureter and ductus deferens. It lies inferior to superior ramus of pubis with companion vein and nerve in canal formed above by bone and below by obturator membrane.

Within the pelvis obturator gives branches to iliac fossa and posterior surface of pubis. These anastomose with pubic branches of inferior epigastric. Obturator artery may arise from inferior epigastric and pass to obturator canal (*abnormal obturator*). If medial to femoral ring on lacunar ligament, in danger in operations for femoral hernia.

Outside pelvis artery divides into anterior and posterior branches which encircle the foramen deep to obturator externus. They supply neighbouring muscles (pectineus, adductors) and hip joint, and anastomose with arteries posterior to hip joint.

Internal pudendal leaves pelvis by greater sciatic foramen below piriformis; winds round ischial spine and enters perineum i.e. below levator ani, by lesser foramen; passes in pudendal canal on medial side of tuberosity of ischium and courses along ischial ramus; perforates the perineal membrane very obliquely.

Relations: In the pelvis, lies lateral to rectum in front of piriformis and sacral plexus. In gluteal region, is deep to gluteus maximus, on ischial spine, at lower border of piriformis, lateral to pudendal nerve and medial to nerve to obturator internus. Thence, with nerve above and vein below, it lies in pudendal canal on lateral wall of ischiorectal fossa medial to obturator internus.

Its *branches* include *inferior rectal* which arises just medial to ischial tuberosity, pierces medial wall of pudendal canal, crosses roof of ischiorectal fossa obliquely and supplies external sphincter and levator ani; *scrotal (labial)* arises in ischiorectal fossa, runs parallel to ischiopubic ramus; crosses superficial transverse perineal muscle into superficial perineal pouch; runs between ischiocavernosus and bulbospongiosus, supplying them and scrotum (labia); *transverse perineal* from scrotal or from trunk near it, runs transversely medially to skin etc.; *artery of bulb* arises near base of perineal membrane which it pierces; also supplies bulbo-urethral gland; *deep artery of penis (clitoris)* pierces perineal membrane to enter crus and is distributed to corpus cavernosum; *dorsal artery of penis (clitoris)* lies between crus and pubic ramus, passes through suspensory ligament, along dorsum of penis (clitoris) with deep dorsal vein medial and dorsal nerve lateral; ends in the glans and prepuce.

Inferior gluteal is terminal branch of anterior trunk (p. 177).

Branches from posterior trunk
Superior gluteal (p. 177).
Iliolumbar passes upwards, backwards and laterally deep to psoas and obturator nerve, but anterior to lumbosacral

trunk; lumbar branch supplies psoas, quadratus lumborum, and gives off a spinal branch through foramen between 5th lumbar vertebra and the sacrum; iliac branch ramifies in iliacus, gives nutrient branch to ilium.

Lateral sacral (2) superior, the larger of the two, to upper part of sacrum and inferior to lower part of sacrum and coccyx; also dorsal branches which enter anterior sacral foramina and end on back of sacrum in muscles and skin.

THE EXTERNAL ILIAC ARTERY

Extends from bifurcation of common iliac at sacro-iliac joint to inguinal ligament midway between symphysis pubis and anterior superior iliac spine.

Relations
Covered by peritoneum and subperitoneal fat; crossed by ureter, ductus deferens and deep circumflex iliac vein; on left by pelvic colon. The gonadal vessels and genitofemoral nerve lie on it for a short distance. Lateral is psoas, except at termination, where it is behind. Medial are its own vein and lymph nodes.

Its *branches* include *inferior epigastric* which arises from front of artery, just above inguinal ligament; passes upwards and medially between peritoneum and fascia transversalis and is crossed laterally by ductus deferens near deep inguinal ring; pierces transversalis fascia, and then passes upwards behind rectus muscle to enter sheath anterior to arcuate line; terminates by anastomosing with superior epigastric branch of internal thoracic. It gives off the *artery to cremaster* which accompanies ductus deferens, and pubic branches which ramify behind pubis and anastomose with pubic of obturator. This artery is frequently enlarged and replaces the obturator artery (*abnormal obturator*).

Deep circumflex iliac arises from lateral side of external iliac near inguinal ligament, is directed to anterior superior iliac spine, then on iliac crest, gradually piercing transversalis fascia and muscle and supplying lower part of abdominal wall.

THE ARTERIES OF THE LOWER LIMB

Inferior gluteal artery

Branch of anterior division of internal iliac; passes out between ventral rami of 1st and 2nd sacral spinal nerves through lower part of greater sciatic foramen, between piriformis and superior gemellus, with sciatic nerve and internal pudendal artery; outside the pelvis it lies between the ischial tuberosity and greater trochanter deep to gluteus maximus, below which it ends in cutaneous branches to back of thigh.

It gives off branches to piriformis, coccygeus and levator ani inside the pelvis.

Outside the pelvis it gives off a coccygeal branch which pierces sacrotuberous ligament and supplies gluteus maximus, integument etc.; a branch to the sciatic nerve; muscular branches to gluteus maximus and lateral rotators of thigh; anastomotic branches (1) to trochanteric fossa (*trochanteric anastomosis*) anastomosing with superior gluteal and ascending branch of medial circumflex, (2) to *cruciate anastomosis* anastomosing with first perforating, medial and lateral circumflex; articular to capsule of hip joint.

Superior gluteal artery

Largest branch of internal iliac, passes laterally between lumbosacral trunk and ventral ramus of 1st sacral nerve, leaves pelvis above piriformis and divides immediately into superficial and deep branches.

Superficial branch runs between glutei, supplying gluteus maximus, and anastomoses with inferior gluteal and medial circumflex. Deep branch goes between gluteus medius and minimus and divides into superior branch which goes to anterior superior iliac spine and anastomoses with circumflex iliacs, ascending branches of lateral circumflex and iliac of iliolumbar, and an inferior branch which supplies gluteal muscles, and descends to greater trochanter. There is also a nutrient branch to ilium.

THE FEMORAL ARTERY

Extends from inguinal ligament to the opening in the adductor magnus where it becomes popliteal artery. With thigh abducted and rotated laterally its course is indicated by the proximal two-thirds of a line drawn from point midway between symphysis pubis and anterior superior iliac spine to adductor tubercle on medial condyle of femur.

Relations
Superficial in upper third of thigh in femoral triangle, more deeply placed in middle third (adductor canal). First 3 cm enclosed in lateral compartment of femoral sheath.

Anterior: Skin, superficial and deep fascia, femoral branch of genitofemoral nerve, sartorius, saphenous nerve, aponeurotic arch over adductor canal.

Posterior: Psoas, profunda vessels, pectineus, adductor longus, femoral vein (at lower part of femoral triangle and in adductor canal), tendon of adductor magnus.

Medial: Femoral vein (in femoral triangle), adductor longus.

Lateral: Femoral nerve; sartorius (in femoral triangle), vastus medialis in adductor canal.

Branches
Three small branches (1) *superficial epigastric* which arises 1 cm distal to inguinal ligament, ascends through the saphenous opening to abdominal wall, (2) *superficial circumflex iliac* which runs laterally to iliac crest, (3) *superficial external pudendal* which pierces cribriform fascia of saphenous opening, runs upwards to pubic tubercle and supplies skin of lower part of abdomen and external genitalia.
Deep external pudendal lies on pectineus, covered by fascia lata, which it pierces; passes medially and is distributed to scrotum (labium).
Profunda femoris arises from lateral and back part of artery about 4 cm distal to inguinal ligament and passes downwards in femoral triangle upon iliacus, psoas, pectineus,

adductor brevis and adductor magnus deep to adductor longus. It is separated from the femoral artery by the femoral and profunda veins and adductor longus. It ends in the lower third of thigh by perforating adductor magnus. It gives off: (a) *Lateral femoral circumflex* (sometimes arises from femoral) passes laterally deep to rectus femoris and sartorius; divides into (1) *transverse branches* which pierce vastus lateralis just below greater trochanter and anastomose below greater trochanter with medial circumflex, first perforating branch of profunda and inferior gluteal (*cruciate anastomosis*), (2) *ascending branches* which pass deep to sartorius, rectus femoris and tensor fasciae latae to anastomose with terminal of superior gluteal and deep circumflex iliac, (3) *descending branch* to knee joint;

(b) *Medial femoral circumflex* which passes backwards between psoas and pectineus, and then below capsule of hip joint between obturator externus and adductor brevis, and divides at the lesser trochanter into two branches; one (ascending) passes upwards along upper border of quadratus femoris to the trochanteric fossa of the greater trochanter, anastomosing with superior and inferior gluteal; the other (transverse) passes to the hamstrings, appears between adjacent borders of quadratus femoris and adductor magnus, anastomoses with first perforating, inferior gluteal and lateral circumflex (*cruciate anastomosis*). An articular branch enters the joint through the acetabular notch;

(c) *Perforating branches* (4) reaching back of thigh by perforating adductor magnus and ending in vastus lateralis; 1st perforates adductor magnus above adductor brevis and is distributed to biceps femoris and gluteus maximus; 2nd perforates adductor brevis and magnus, is distributed to hamstrings and gives off a nutrient artery to femur; 3rd arises at the lower border of adductor brevis and perforates magnus; 4th is the name applied to terminal part of profunda, which pierces adductor magnus near opening for femoral vessels and supplies short head of biceps femoris.

The femoral artery gives off muscular branches and the *descending genicular* which arises in adductor canal. Divides into superficial branch (*saphenous artery*) which runs with saphenous nerve, and deep branch which joins anastomosis around knee joint.

THE POPLITEAL ARTERY

Extends from opening in adductor magnus to lower border of popliteus where it divides into anterior and posterior tibial. It is about 20 cm long. Its upper part inclines from medial side of femur to middle of intercondylar notch. From there it lies in midline of popliteal space.

Relations
The upper part of artery in popliteal space is overlapped medially by semimembranosus, and inferiorly is covered by gastrocnemius and plantaris. It rests upon the femur, oblique popliteal ligament of knee joint and popliteus. The vein is superficial to the artery, somewhat lateral superiorly and medial inferiorly. The tibial (medial popliteal) nerve is superficial and slightly lateral to both vessels superiorly but inferiorly the nerve crosses to the medial side. A small branch of the obturator nerve courses upon the artery.

Branches
Muscular to hamstrings, both heads of gastrocnemius, plantaris and soleus.
Genicular—lateral and medial superior, and lateral and medial inferior. Run horizontally deep to muscles and ligaments and take part in anastomosis round knee joint. Anastomosis includes descending genicular, descending branch of lateral femoral circumflex, 4th perforating, and recurrent branches of anterior tibial and circumflex fibular from posterior tibial.
Middle genicular pierces oblique popliteal ligament and supplies structures in joint.

THE ANTERIOR TIBIAL ARTERY

Extends from division of popliteal artery at lower border of popliteus to front of ankle, where it becomes the dorsalis pedis artery. It passes forwards between two heads of origin of tibialis posterior through interosseous membrane to reach its anterior surface; thence it is indicated by a line drawn from medial side of head of fibula to midway between malleoli.

Relations
Lies deeply on interosseous membrane; tibialis anterior to medial side and extensor digitorum longus above, and extensor hallucis longus below on lateral side; covered inferiorly by extensor retinacula and crossed from lateral to medial by extensor hallucis longus tendon; rests distally on the anterior surface of the tibia. Accompanied by two veins. Deep peroneal (anterior tibial) nerve lies at first lateral, then superficial, then again lateral.

Branches
Posterior tibial recurrent arises before artery perforates interosseous membrane and passes upwards deep to popliteus to back of knee. *Circumflex fibular* passes round neck of fibula, through soleus, to peroneus longus. *Anterior recurrent* arises as artery reaches anterior surface of interosseous membrane and passes in tibialis anterior to lateral and anterior surfaces of knee joint. *Lateral and medial anterior malleolar* arise just above ankle joint and supply it. *Muscular* to surrounding muscles.

Dorsalis pedis artey

Extends from front of ankle to proximal part of 1st metatarsal space and ends by entering sole between heads of 1st dorsal interosseous muscle and anastomosing with lateral plantar artery to form plantar arch.

Relations
Lies between tendons of extensor hallucis longus medially and extensor digitorum longus laterally; near termination crossed by tendon of extensor hallucis brevis. Bound down by fascia as it lies on talus, navicular and intermediate cuneiform. Accompanied by two veins. Deep peroneal nerve is lateral.

Branches
Tarsal arises as artery crosses navicular and supplies extensor digitorum brevis. *Arcuate* arises near bases of metatarsals and arches laterally over bases of metatarsal bones deep to short extensor of toes to lateral side of foot. From

the convexity of arch *three dorsal metatarsal* arteries pass to three lateral metatarsal spaces. They supply the interossei and divide at cleft of toes into *digital branches*; the most lateral one also supplies lateral side of little toe. Each dorsal metatarsal artery communicates at the cleft of the toes with a plantar digital by an *anterior perforating branch*, and at the back of the metatarsal space with the plantar arch by a *posterior perforating branch*.

1st dorsal metatarsal arises as dorsalis pedis is about to dip down into sole. It lies over dorsum of 1st metatarsal space, and divides at cleft to supply contiguous sides of 1st and 2nd toes, having previously given off a branch to medial side of great toe.

1st plantar metatarsal (arteria magna hallucis) arises in sole from dorsalis pedis at junction with plantar arch; passes forwards in 1st metatarsal space to cleft, where it divides into two branches for contiguous sides of 1st and 2nd toes, having previously given off a branch to medial side of great toe.

THE POSTERIOR TIBIAL ARTERY

Extends from lower border of popliteus to lower edge of flexor retinaculum of ankle, where it divides into medial and lateral plantar midway between medial malleolus and medial border of heel. At first midway between tibia and fibula, then approaches tibia and lies on it.

Relations

Upper two-thirds covered by gastrocnemius and soleus. Lower one-third superficial between medial border of tendo calcaneus and medial border of tibia. Tibial nerve is at first on medial side, but 5 cm below origin nerve crosses posteriorly to reach lateral side. Has two veins and lies on tibialis posterior, flexor digitorum longus, tibia, and back of ankle joint.

At medial malleolus, from medial to lateral there are tendons of tibialis posterior and flexor digitorum longus, vein, artery, vein, nerve, tendon of flexor hallucis longus.

Branches
Peroneal arises 2.5 cm from popliteus, courses obliquely to fibula, then along medial border of that bone, between origins of tibialis posterior and flexor hallucis longus to lower part of interosseous membrane, where it gives off a perforating branch. It continues over inferior tibiofibular articulation to lateral side of lateral malleolus where it ends by anastomosing with lateral plantar and tarsal arteries. Covered in upper part by soleus and deep fascia, then by flexor hallucis longus. Beyond the malleolus it is superficial.

The peroneal artery gives off *muscular* branches to soleus, tibialis posterior, flexor hallucis longus and peronei, a *nutrient* branch to the fibula, a *perforating* branch which arises about 5 cm above lateral malleolus, pierces or passes below interosseous membrane, and under cover of peroneus tertius reaches front of lateral malleolus and tarsus, supplying ankle joint and anastomosing with lateral malleolar, tarsal of dorsalis pedis and terminal of peroneal.

The posterior tibial also gives off *muscular* branches to muscles of back of leg, a *nutrient* to tibia which arises near origin of posterior tibial (largest of its kind in body), a *communicating* branch which arises 5 cm above medial malleolus, passes deep to flexor hallucis longus and anastomoses with communicating of peroneal and a *calcaneal* which arises near termination, pierces flexor retinaculum with medial calcanean nerve and supplies skin, fat of heel and muscles on medial side of foot.

Medial plantar artery

Smaller terminal branch of posterior tibial directed forwards along medial border of foot as far as base of 1st metatarsal bone, thence along medial side of great toe to anastomose with 1st plantar metatarsal. It is covered at first by abductor hallucis, and subsequently becomes more superficial, lying between that muscle and flexor digitorum brevis; it has medial plantar nerve lateral to it.

It gives off digital branches to clefts between three medial toes which join digital branches from plantar arch and also branches to skin and muscles of medial side of sole of foot.

Lateral plantary artery

From medial part of foot deep to abductor hallucis it runs laterally with lateral plantar nerve, between flexor digitorum brevis and flexor digitorum accessorius, to base of 5th metatarsal; thence it passes medially, resting on the interosseous muscles and deep to long flexor tendons and lumbricals to posterior part of 1st metatarsal space, and anastomoses with dorsalis pedis, completing plantar arch.

The plantar arch lies at level of bases of metatarsals and is accompanied by deep branch of the lateral plantar nerve.

The plantar arch gives off *recurrent branches* which pass backwards to tarsal joints, *posterior perforating branches* (3) which ascend to dorsum of foot through posterior part of three lateral metatarsal spaces to anastomose with dorsal metatarsals of arcuate, and *plantar metatarsal* (4) which supply both sides of three lateral toes and lateral half of 2nd (*plantar digital*); medial three bifurcate at the cleft of toes, also give off at point of division *anterior perforating* to anastomose with dorsal metatarsal arteries.

4
The Veins

THE VEINS OF THE HEAD AND NECK

The veins of the brain

These veins have no valves and owing to the absence of muscular tissue have very thin walls.

Cerebral veins may be divided into *superficial*, which lie in sulci on the surface of the hemisphere; and *deep*, which issue from the substance of the brain.

The *superficial veins* are divisible into three sets: (a) *superior* (10 to 12 on each side), which pass forwards and upwards to superior sagittal sinus; (b) *inferior*, of which three are large *superficial middle cerebral*, along lateral cerebral fissure to cavernous sinus, often connected with superior sagittal sinus by *superior anastomotic vein*, and transverse sinus through *inferior anastomotic vein*; (c) *medial, deep middle cerebral vein* on insula, joined by *anterior cerebral* and *striate veins* at anterior perforated substance to form *basal vein*. Basal vein winds around cerebral peduncle, collecting thalamic veins from posterior perforated substance, meets *great cerebral vein* and *inferior sagittal sinus* at commencement of *straight sinus*.

The *deep veins* are the *choroid vein*, which drains choroid plexus of lateral ventricle, unites with veins of septum lucidum and *thalamostriate* to form *internal cerebral vein* near interventricular foramen; the right and left internal cerebral veins run backwards between layers of tela choroidea of 3rd ventricle, below splenium of corpus callosum, to form *great cerebral vein*, which emerges from transverse cerebral fissure to enter anterior end of straight sinus.

Cerebellar veins are superior and inferior; the former open into straight, the latter into transverse and superior petrosal sinuses.

The cerebral venous sinuses (Figs. 45 and 46)

Their structure is like that of veins of brain. They are situated between two layers of dura mater.

185

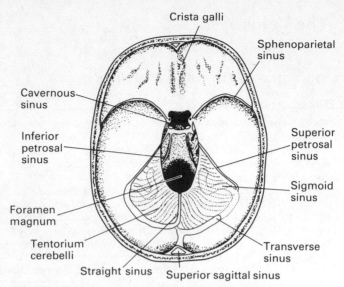

Fig. 45 Intracranial venous sinuses (from above).

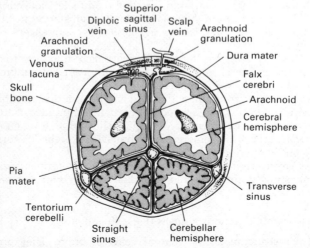

Fig. 46 Intracranial venous sinuses (coronal section posteriorly).

Superior sagittal begins above crista galli, runs backwards in upper border of falx cerebri to *confluence of sinuses*, usually forming *right transverse sinus*; receives superior cerebral, parietal and emissary veins. The lumen is triangular and is intersected by fibrous trabeculae. On either side open the *lacunae laterales*. *Arachnoid granulations* and *villi* project into lumen.

Inferior sagittal sinus runs along posterior two-thirds of inferior margin of falx cerebri to straight sinus. Receives blood from medial surface of hemispheres.

Straight sinus lies at junction of tentorium and falx cerebri, goes to confluence of sinuses, and usually forms *left transverse sinus*; receives inferior sagittal sinus, great cerebral, basal and superior cerebellar veins.

Transverse sinus, one to each side, runs from confluence of sinuses to lateral angle of occipital bone between layers of tentorium, then as *sigmoid sinus* curves down on mastoid part of temporal bone to jugular foramen where it becomes *internal jugular vein*; each receives superior petrosal sinus and mastoid vein; inferior petrosal sinus joins internal jugular just below jugular foramen; the right transverse is usually formed by superior sagittal and the left by the straight sinus.

Occipital sinus is small and runs from confluence of sinuses in falx cerebelli, divides around foramen magnum and enters termination of sigmoid sinus.

Cavernous sinuses one on each side of sella turcica, extend from superior orbital fissure to apex of petrous part of temporal. Receive ophthalmic veins which connect the anterior facial vein with this sinus; also superficial middle cerebral veins and sphenoparietal sinus. The *ophthalmic veins* are two in number, *superior* and *inferior*, and pass from orbit to cranial cavity through the superior orbital fissure. They communicate through inferior orbital fissure with pterygoid plexus.

Cavernous sinus broken into multiple small venous channels by honeycomb of fibrous septa. Through the sinus run internal carotid artery (S-bend), sympathetic plexus and abducent nerve. In lateral wall lie oculomotor, trochlear and ophthalmic and maxillary divisions of trigeminal nerve.

Intercavernous sinus surrounds pituitary gland and connects the cavernous sinuses.

Inferior petrosal sinus, one on each side, from termination of cavernous to internal jugular vein; lies over petro-occipital suture.

Basilar sinuses form a network which connects the inferior petrosal sinuses and lie on basilar part of occipital bone.

Superior petrosal sinus, one on each side, is on superior border of petrous part of temporal and connects transverse and cavernous sinuses; receives inferior lateral cerebral, labyrinthine (internal auditory) and anterior lateral cerebellar veins.

Sphenoparietal sinus runs medially along free margin of lesser wing of sphenoid to cavernous sinus.

The diploic veins

Lodged in channels between inner and outer layers of bone forming the cranial vault. They are divided into *frontal*, joining the supra-orbital vein; *anterior parietal*, joining a deep temporal vein; *posterior parietal*, joining transverse sinus; and *occipital*, joining occipital vein (outside) or transverse sinus (inside skull).

The emissary veins

Small veins passing through foramina in the bones and connecting the sinuses inside the skull with the external veins of the head.

THE VEINS OF THE FACE, SCALP AND NECK (Fig. 47)

Facial (anterior facial): Passes obliquely across side of face from medial angle of orbit to anterior border of masseter posterior to artery; at origin connected with supratrochlear and supra-orbital veins and through superior ophthalmic veins with cavernous sinus. Crossing the mandible, it unites with the anterior division of the *retromandibular (posterior facial)* vein to form a short trunk (*common facial*), which crosses submandibular gland, digastric and external

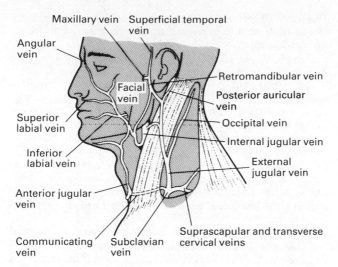

Maxillary vein Superficial temporal vein

Angular vein

Facial vein

Retromandibular vein

Posterior auricular vein

Occipital vein

Internal jugular vein

External jugular vein

Superior labial vein

Inferior labial vein

Anterior jugular vein

Communicating vein   Subclavian vein   Suprascapular and transverse cervical veins

Fig. 47 Superficial veins of head and neck.

carotid artery superficially and joins the internal jugular vein.

The facial vein receives supra-orbital, supratrochlear, tributaries from eyelids, external nose, lips, glands (parotid and submandibular), cheek and masseter, submental and ascending palatine. It is connected to the pterygoid plexus by the deep facial veins. This plexus and the supra-orbital through the ophthalmic connect the facial vein with the cavernous sinus.

Superficial temporal comes from side and vertex of head, passes down over zygoma, where it receives the middle temporal vein; it then passes downwards between condyle of mandible and external acoustic meatus into the substance of the parotid, where it joins the maxillary vein to form the retromandibular (posterior facial) vein. The temporal vein receives tributaries from the temporal region, parotid gland and auricle and the transverse facial vein.

Maxillary is formed by tributaries corresponding with the branches of the maxillary artery; they form the *pterygoid plexus* placed around and within lateral pterygoid muscle.

The trunk of the vein passes backwards with artery to neck of mandible to join *superficial temporal vein* and form *retromandibular (posterior facial) vein*.

The pterygoid plexus receives tributaries corresponding with branches of maxillary artery (p. 151).

The pterygoid plexus communicates with inferior ophthalmic vein, with the facial vein by the deep facial vein, and with cavernous sinus via emissary veins passing through foramen ovale or foramen of Vesalius.

Retromandibular (posterior facial) vein is formed by union of superficial temporal and maxillary veins; descends in parotid gland superficial to external carotid artery and deep to facial nerve; divides into two branches—anterior, which unites with the (anterior) facial, and posterior, which receives the posterior auricular vein to form the external jugular vein.

Posterior auricular from plexus on side of head and back of ear, receives stylomastoid vein, and branches from external ear; joins with posterior division of retromandibular to form external jugular vein. Communicates with sigmoid sinus by mastoid emissary vein.

Occipital from plexus in back part of scalp; it then passes deeply under semispinalis capitis, over suboccipital triangle, where it communicates with vertebral, and joins deep cervical vein. It may follow occipital artery and join internal jugular vein. May join posterior auricular.

External jugular is formed by junction of posterior division of retromandibular trunk and posterior auricular vein at angle of mandible in the substance of parotid. Descends deep to platysma over sternocleidomastoid, pierces cervical fascia above middle of clavicle to open into subclavian (occasionally into the internal jugular). Has two pairs of valves. Receives posterior external jugular, draining superficial region at back of neck, suprascapular, transverse cervical and anterior jugular.

Anterior jugular drains skin and superficial muscles of submental region; passes downwards on either side of midline and reaches suprasternal notch; communicates with fellow (*jugular arch*) and passes laterally deep to sternocleidomastoid to end in external jugular.

Internal jugular from jugular foramen is continuation of

sigmoid sinus. Passes vertically down the side of neck, lateral to internal and common carotid arteries (p. 147), and unites with subclavian near the medial margin of the scalenus anterior to form the brachiocephalic vein. Usually crossed by accessory nerve and inferior root of ansa cervicalis. Has a pair of valves 1.5 cm above termination.

The internal jugular vein receives inferior petrosal sinus, pharyngeal, lingual (dorsal and deep veins of tongue and vena comitans of hypoglossal nerve), (common) facial and superior and middle thyroid veins.

Vertebral: There is no vertebral vein inside the skull. Communicates with posterior spinal and occipital veins; starts in occipital region and drains deep muscles of back of neck; enters foramen in transverse process of atlas, runs down in front of artery through foramina transversaria of the cervical vertebrae to 6th (or 7th), whence it passes downwards to enter brachiocephalic vein. One pair of valves guard it at its origin.

The vertebral vein receives muscular, spinal, deep cervical and 1st intercostal veins.

Deep cervical corresponds with deep cervical artery; lies deeply in neck; receives occipital superiorly and ends in vertebral.

THE VEINS OF THE UPPER LIMB

(a) *Superficial*, in superficial fascia on deep fascia.

Dorsal venous network: Back of hand; receives bulk of blood from palm, whence pressure of gripping drives it; medial side drains into basilic and lateral side into cephalic veins (more into former); receives *digital veins* via *metacarpal veins.*

Cephalic from dorsal venous network on lateral side of forearm passing more anteriorly towards elbow; lateral side of upper arm lateral to biceps brachii, between deltoid and pectoralis major, pierces clavipectoral fascia and ends in axillary vein. In front of elbow gives off large *median cubital vein* passing medially to join basilic vein.

Basilic from dorsal venous arch on medial side of forearm passing more anteriorly towards elbow; receives median cubital and continues upwards medial to biceps brachii; about middle of upper arm pierces deep fascia, runs upwards medial to brachial artery and becomes *axillary vein* at lower border of teres major; usually receives brachial vein(s).

Median vein of forearm from superficial veins of palmar surface of hand; runs up middle of forearm, communicating below bend of elbow with ulnar veins; then joins median cubital or basilic; may divide into median cephalic and median basilic, former to cephalic, latter to basilic.

(b) *Deep* (these accompany the main arteries and are usually paired, *venae comitantes*; they also communicate with the superficial veins).

Each digit has two *digital veins* which drain into the *superficial palmar veins*. These drain into the *radial* and *ulnar veins*. *Deep palmar vein* goes to the radial. Radial and ulnar veins join to form the *brachial veins* which after receiving tributaries corresponding to branches of brachial artery become the *axillary vein* at lower border of teres major. Basilic vein joins brachial about level of middle of upper arm. Axillary receives tributaries corresponding to branches of axillary artery and also cephalic vein before becoming *subclavian vein* at outer border of 1st rib. There are valves near beginning of axillary vein at junction with cephalic and subscapular veins.

Subclavian vein, in front of and at lower level than artery, from which separated by scalenus anterior and phrenic nerve; ends behind medial end of clavicle by joining *internal jugular* to form *brachiocephalic*. Most of veins corresponding to branches of subclavian artery join vertebral or brachiocephalic vein or external jugular (p. 140) which joins subclavian vein lateral to sternocleidomastoid. There is a valve in the subclavian vein about 2.5 cm lateral to termination before joined by external jugular.

THE VEINS OF THE THORAX

Brachiocephalic (innominate): Two large trunks, formed by junction of internal jugular and subclavian veins of corresponding side; end by uniting to form superior vena cava; no valves.

The *right*, 2.5 cm long, is formed behind medial end of right clavicle; passes downwards to join left brachiocephalic at the inferior border of 1st right costal cartilage. Receives the right vertebral, right internal thoracic, right inferior thyroid and 1st right intercostal veins. Right lymphatic duct opens into it at the angle of union of right subclavian and internal jugular veins.

The *left*, 7.5 cm long, passes from left to right and downwards just above arch of aorta, superficial to its branches and phrenic and vagus nerves. Thoracic duct opens into it at angle of union of left subclavian and internal jugular veins.

The left brachiocephalic vein receives left vertebral, left internal thoracic, left inferior thyroid, first left intercostal, left superior intercostal, pericardiacophrenic, tributaries from mediastinum and thymus.

Internal thoracic, two with each artery, unite in a single trunk; end in brachiocephalic vein. Receives tributaries corresponding with branches of internal thoracic artery.

Inferior thyroid (sometimes three or four) from thyroid venous plexus usually join left brachiocephalic vein.

Intercostal, eleven on each side, run above corresponding arteries in intercostal space. The *1st intercostal* passes up over neck of 1st rib and over pleura to end in brachiocephalic.

The right second and third intercostal veins join to form *right superior intercostal*, which ends in azygos.

The corresponding vein on the left receives left bronchial vein and passes forward across arch of aorta to left brachiocephalic vein. (For other intercostal veins see vena azygos.)

Superior vena cava, 7.5 cm long, is formed by union of right and left brachiocephalic veins behind the junction of 1st right costal cartilage with sternum; passes downwards to join right atrium opposite upper border of 3rd right costal cartilage. Superior vena cava enters the pericardium about 3 cm from its termination, and this part is covered with serous pericardium except posteriorly. No valves. Receives pericardiac and mediastinal veins and, just as it enters the pericardium, the azygos vein.

Azygos vein commences to right of 1st or 2nd lumbar vertebra as a branch from right lumbar veins, passes up through aortic opening in diaphragm to right of aorta and along right side of vertebral column in front of right intercostal arteries to 5th thoracic vertebra, where, arching over root of right lung, it joins superior vena cava. Receives the eight lower right intercostal veins, right superior intercostal vein, hemiazygos veins and oesophageal, mediastinal and right bronchial veins. Imperfect valves, though its tributaries have complete ones.

Hemiazygos vein commences in lumbar region of left side from lumbar veins, or branches of renal, passes through left crus of diaphragm to level of 8th thoracic vertebra, where it crosses behind aorta, thoracic duct and oesophagus to terminate in vena azygos. Receives three or four lower left intercostal and some oesophageal and mediastinal veins.

Accessory hemiazygos vein is formed by union of 4th, 5th, 6th, 7th and 8th left intercostal veins; communicates above with superior intercostal vein; crosses 7th thoracic vertebra behind aorta, thoracic duct and oesophagus to end in vena azygos.

Bronchial veins from lungs; the right terminate in azygos and the left in the left superior intercostal or hemiazygos.

Pulmonary veins (two on each side) commence in capillary network in lung and unite to form a trunk for each lobe; the vein of the middle lobe of the right lung unites with that of the superior lobe; hence there are two veins on each side. No valves. Join left atrium; left veins cross anterior to the descending thoracic aorta; right pass behind right atrium and inferior vena cava.

Cardiac: (p. 139).

THE VERTEBRAL AND SPINAL VEINS

External vertebral plexuses extend along whole length of vertebral column forming network on bodies of vertebrae in front and on vertebral arches behind; terminate in vertebral (neck), intercostal (thorax), lumbar (abdomen) and sacral (pelvis) veins.

Internal vertebral plexuses are situated between vertebrae and dura mater and consist of anterior and posterior longitudinal vertebral plexuses which run whole length of vertebral canal; both sets terminate in external plexuses. Plexuses form communication betwee pelvic veins and intracranial sinuses and veins.

Basivertebral veins emerge from back of bodies of vertebrae and terminate in anterior longitudinal; very large, since they drain active red marrow.

Veins of the spinal cord cover whole length of cord, between pia and arachnoid, communicate along nerve roots with segmental veins (vertebral, intercostal, lumbar, sacral); no valves in any of the spinal veins.

THE VEINS OF THE ABDOMEN AND PELVIS

Inferior vena cava (Fig. 48) is formed by union of the two common iliac veins a little to the right in front of 5th lumbar vertebra behind right common iliac artery; passes upwards

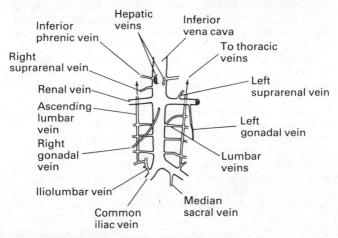

Fig. 48 Inferior vena cava.

on right side of aorta to inferoposterior surface of liver, where it becomes embedded in a groove and receives the hepatic veins; thence it goes through vena caval opening in diaphragm between middle and right tendinous leaflets at level of 8th thoracic vertebra and 6th right costal cartilage, enters pericardium and opens into lower and back part of right atrium; its orifice has an imperfect valve.

Relations
The inferior vena cava lies behind the superior and horizontal parts of the duodenum and the head of the pancreas. Above the duodenum, the epiploic foramen with the portal vein in the lesser omentum is anterior. The mesentery crosses anteriorly and the right gonadal, right colic and right common iliac arteries are also anterior. Posteriorly the bodies of the lumbar vertebrae with the right sympathetic trunk lie along its whole length. The right crus, psoas and suprarenal gland are posterior near the diaphragm. The right renal, lumbar, suprarenal and inferior phrenic arteries are posterior. The abdominal aorta lies to its left.

Tributaries: In addition to the common iliac and hepatic veins the inferior vena cava receives lumbar, right gonadal, right and left renal, right suprarenal and inferior phrenic veins.

Lumbar: Usually five; upper two end in azygos or hemiazygos, 3rd and 4th in inferior vena cava and 5th in iliolumbar. Joined by vertically running *ascending lumbar vein* ending on right in azygos and on left in hemiazygos.

Gonadal: Testicular veins emerge from deep inguinal ring and drain pampiniform plexus of spermatic cord. Form one vein which opens into inferior vena cava on right and left renal vein on left. *Ovarian veins* from pampiniform plexus in broad ligament draining ovary. Accompany ovarian artery and form one vein ending in same manner as testicular.

Renal: Right about 2.5 cm long lies in front of renal artery behind descending part of duodenum; *left* about 7.5 cm long also in front of artery; passes to right in front of aorta, behind superior mesenteric artery and body of pancreas and above horizontal part of duodenum. Receives left gonadal and left suprarenal veins.

Suprarenal: Single vein from each gland; short right joins inferior vena cava; longer left joins left renal vein.

Common iliac extends from brim of pelvis near sacro-iliac joint where formed by union of internal and external iliac veins to front of body of 5th lumbar vertebra, a little to the right of the middle line behind right common iliac artery, where with the left common iliac it forms inferior vena cava. The right vein is shorter, and nearly vertical. Receives iliolumbar and sometimes lateral sacral veins. Middle sacral joins left common iliac. No valves.

Relations
Right vein passes behind, and then to right side of artery; it is anterior to obturator nerve and iliolumbar artery. Left vein is placed on medial side of left artery, and then passes behind right common iliac artery to join right vein; crosses middle sacral artery and is crossed by superior rectal vessels.

Internal iliac is formed by the union of all of the veins of the branches of the internal iliac artery, except the iliolumbar veins which open into the common iliac. It lies at first on the medial side of and then behind the artery. It has no valves. The visceral veins opening into the internal iliac anastomose very freely and come from a series of plexuses—rectal, vesical, prostatic, uterine, vaginal. The submucosal rectal plexus drains upwards on the whole to the superior rectal vein which joins the inferior mesenteric. The middle rectal vein joins the internal iliac, and the inferior rectal, the internal pudendal. Prostatic plexus drains mainly into the vesical which forms vesical veins going to the internal iliac. Prostatic plexus receives deep dorsal vein of penis. Vesical receives dorsal vein of clitoris. There are vaginal and uterine veins draining the plexuses and going to the internal iliac.

The parietal veins joining the internal iliac are lateral sacral, superior and inferior gluteal (from the buttock entering the pelvis through the greater sciatic foramen), obturator (entering through obturator canal), internal pudendal (corresponding to artery, p. 175).

External iliac, continuation of femoral; begins deep to inguinal ligament and continues along brim of pelvis to sacro-iliac joint where unites with internal iliac to form common iliac. On right side lies medial to artery at first, but gradually passes behind it. On left side, medial to artery. Receives inferior epigastric and deep circumflex iliac, and a pubic vein from the obturator.

THE HEPATIC PORTAL SYSTEM OF VEINS (Fig. 49)

These veins, which collect the blood from the digestive tract, are valveless. They form a trunk, the *portal vein*, which enters the liver and breaks up into small branches in

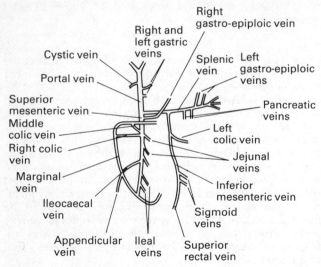

Fig. 49 Portal venous system.

its substance. Thus the blood, having passed through capillaries in the bowel wall, again passes through capillary-like vessels, termed *sinusoids*, in the liver. Oxygen to liver substance is supplied by hepatic artery, and blood from both hepatic artery and portal vein goes to inferior vena cava by *hepatic veins*. The following veins form the portal system.

Inferior mesenteric drains upper two-thirds of anal canal, rectum, pelvic colon, descending colon and left half of transverse colon. It lies to the left of its artery, and passes near the duodenojejunal flexure, in anterior fold of para-duodenal fossa. Then, passing behind pancreas and in front of left kidney, opens into the splenic vein. Receives *superior* and *inferior left colic* and *superior rectal veins*. (The rectal tributaries anastomose with the rectal tributaries of the internal iliac forming a portal–systemic anastomosis.)

Superior mesenteric drains small intestine (beyond middle of descending part of duodenum), caecum, ascending and right half of transverse colon. It passes upwards in front and to right of superior mesenteric artery, in front of horizontal part of duodenum and behind neck of pancreas; joins the splenic vein behind the upper border of the pancreas to form the portal vein. Receives jejunal, ileal, right gastro-epiploic, middle colic, inferior pancreatico-duodenal, right colic and ileocolic veins.

Splenic, commences in five or six tributaries in hilum of spleen; these unite to form a trunk which passes inferior to splenic artery from left to right, behind pancreas, and in front of superior mesenteric artery; joins superior mesenteric vein at a right angle. Receives pancreatic, short gastric, left gastro-epiploic and inferior mesenteric veins.

Left gastric is a large vein accompanying left gastric artery from right to left along lesser curvature of stomach to cardia, where it receives oesophageal tributaries and passing to right behind lesser sac, opens into the portal vein. Oesophageal veins near cardia constitute another portal–systemic anastomosis.

Portal vein is formed by the union of the splenic and superior mesenteric veins in front of the right crus of diaphragm and inferior vena cava, and behind the neck of the pancreas. Passes up behind superior part of duodenum and then between the layers of the lesser omentum, behind and between the bile duct and hepatic artery, the duct being on the right and artery on the left, to porta hepatis. Here it divides into right and left branches to lobes of liver. Left supplies quadrate and most of caudate lobes as well as anatomical left lobe. Connected with the branch to the left lobe are the *ligamentum teres* of liver (fetal obliterated left umbilical vein) in front and the *ligamentum venosum* (fetal ductus venosus) behind. This constituted the means whereby the blood from the placenta bypassed the liver.

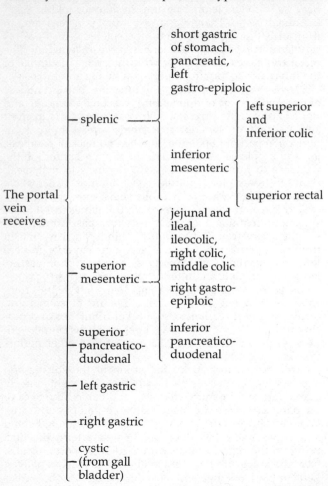

The portal vein receives

- splenic
  - short gastric of stomach, pancreatic, left gastro-epiploic
  - inferior mesenteric
    - left superior and inferior colic
    - superior rectal
- superior mesenteric
  - jejunal and ileal, ileocolic, right colic, middle colic
  - right gastro-epiploic
- superior pancreatico-duodenal
  - inferior pancreatico-duodenal
- left gastric
- right gastric
- cystic (from gall bladder)

THE VEINS OF THE LOWER LIMB

(a) *Superficial* (in superficial fascia on deep fascia).

Dorsal venous arch, proximal to heads of metatarsals, receives most of blood from sole of foot.

Great saphenous arises from medial side of arch on dorsum of foot; ascends in front of medial malleolus, accompanied by saphenous nerve; bends behind medial condyle of femur; ascends along medial side of thigh to 3 cm inferior and lateral to pubic tubercle where it receives superficial circumflex iliac, superficial epigastric and superficial external pudendal veins; then passes through saphenous opening in the deep fascia and joins the femoral. Communicates with deep veins, especially in soleus muscle. Through these, and other communicating veins above the knee, blood flows from superficial to deep veins, due to valves; also many valves along its length.

Small saphenous arises from lateral side of plexus on dorsum and passes upwards behind lateral malleolus to median line of calf, accompanied by sural nerve; pierces deep fascia on back of knee and joins the popliteal vein between the heads of the gastrocnemius; has a valve near termination. Communicates with deep veins of calf and with great saphenous.

(b) *Deep* (these accompany the main arteries and the smaller are paired, venae comitantes).

Posterior tibial veins formed from lateral and medial plantar, joining with the peroneal. Course same as artery (p. 182).

Anterior tibial veins, continuation of dorsal veins of foot; pierce interosseous membrane in upper part of leg, and form the popliteal vein, by junction with the posterior tibial veins, at the lower border of the popliteus muscle.

Popliteal vein passes upwards to femoral aperture in adductor magnus where it becomes the femoral; receives sural and genicular veins; has four valves. Vein is superficial to artery which it crosses from medial to lateral.

Femoral vein passes from the opening in the adductor magnus to inguinal ligament where it becomes external iliac. Lies at first lateral to artery, but higher up crosses behind to its medial side. Receives muscular branches, profunda femoris and great saphenous. Four or five valves. Lies in middle compartment of femoral sheath medial to femoral canal which provides space for expansion of vein during increased blood flow.

5
The Lymphatic System, Spleen and Thymus

The *thoracic duct* (Fig. 50) receives the lymphatic vessels from both lower limbs, abdomen, except upper surface of liver, from left half of thorax, left upper limb, and left side of head and neck. It is about 40–45 cm long and about 5 mm wide and extends from right of body of 2nd lumbar vertebra, where it commences as a dilatation, the *cisterna chyli*, to junction of left internal jugular with left subclavian vein. Contains numerous valves.

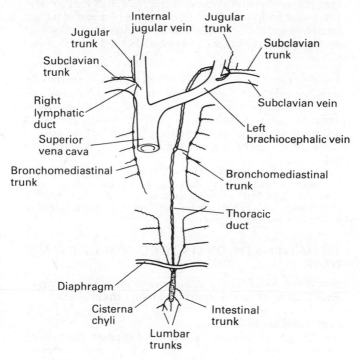

Fig. 50   Main lymph trunks and ducts.

Relations: The abdominal part lies on the front of the body of the 2nd lumbar vertebra, behind and to the right side of the aorta, on the medial side of the right crus; it enters the thorax through aortic opening, on the right side of aorta, lying between it and the vena azygos, and passes upwards to right of aorta anterior to right intercostal arteries and terminations of hemiazygos veins. Crosses to left at about level of 5th thoracic vertebra behind oesophagus. About 4th thoracic vertebra it lies to the left behind aortic arch and runs along the left side of the oesophagus behind the left common carotid artery. At level of 7th cervical vertebra it turns laterally, passes behind left internal jugular vein and in front of 1st part of left subclavian artery and arches over apex of left pleura where it receives left jugular and subclavian lymphatic trunks; joins angle of union of the left internal jugular and left subclavian veins. May break into several terminal branches. Appears beaded due to many valves.

The *right lymphatic duct* receives the lymphatic vessels of the right upper limb, right side of thorax, right half of head and neck, and upper surface of liver. Usually formed by union of right jugular and subclavian lymphatic trunks. It is about 1–2 cm long and joins the venous system at the angle of union of the right internal jugular and right subclavian veins.

Note: Lymphatic vessels communicate freely with veins throughout body so that thoracic duct can be ligated without harm. Superficial lymphatic vessels follow veins, deep follow arteries. Lymphatic vessels richly supplied with valves so that pulsation of nearby arteries aids lymph flow.

THE LYMPHATIC DRAINAGE OF THE HEAD AND NECK

Superficial nodes form a collar at level of chin and occipital region; deep nodes lie vertically along carotid sheath.

Superficial nodes

Suboccipital nodes (1 or 2) at level of superior nuchal line receive lymphatic vessels from back of scalp; efferent vessels join superficial cervical nodes.

Retro-auricular (mastoid) nodes (2 or 3) behind auricle receive vessels from back of auricle and external acoustic meatus; efferent vessels join superficial cervical nodes.

Parotid nodes (3 or 4) (one lies just anterior to tragus) receive lymphatics from temporal region, external acoustic meatus and lateral parts of eyelids; efferent vessels pass to submandibular and superficial cervical nodes.

Facial (buccal) nodes (2–3 small) along facial vein receive afferents from medial parts of eyelids, cheek, lateral parts of lips, external nose; efferents to submandibular.

Submental nodes receive vessels from tip of tongue, gums of mandibular incisors and middle of floor of mouth and lower lip; efferent vessels mainly to submandibular lymph nodes.

Submandibular nodes (8 to 10) afferent vessels from face, medial parts of eyelids, tongue, teeth, gums, front of nasal cavity and sinuses, submandibular and sublingual glands; efferent vessels to cervical nodes.

Superficial cervical nodes (4 to 6) lie along (a) external jugular vein, (b) vertically near midline of neck, (c) along line of accessory nerve in posterior triangle. Afferent vessels from external ear, skin of neck; efferent vessels to deep cervical nodes.

Deep nodes

Deep cervical nodes (20 to 30) are divided into superior and inferior.

Superior along internal jugular vein from division of common carotid to base of skull. *Jugulodigastric nodes* near digastric especially associated with afferents from palatine tonsil. Afferent vessels from submandibular nodes, cranium, tongue, larynx, lower part of pharynx, and thyroid gland. Efferent vessels to inferior group.

Inferior placed along lower part of internal jugular vein. *Jugulo-omohyoid node* near omohyoid especially associated with afferents from tongue. Afferent vessels from other cervical nodes and lower part of neck. Efferent vessels form a single trunk (*jugular lymphatic trunk*), opening into thoracic duct on left side, and into the right lymphatic duct on right side.

Retropharyngeal nodes behind nasopharynx in front of vertebral column receive afferents from pharyngeal tonsil (adenoid) and drain to superior deep cervical nodes.

Prelaryngeal, pretracheal and paratracheal nodes, in front and at sides of larynx and trachea, receive afferents from thyroid gland, lower anterior half of larynx and whole of trachea; efferents to inferior deep cervical.

Lymphatic drainage of the tongue

Tip drains to submental nodes; sides homolaterally to submandibular nodes; midline bilaterally to submandibular; posterior third bilaterally to jugulodigastric and, especially, jugulo-omohyoid directly.

Lymphatic drainage of the thyroid gland

To nodes near larynx and trachea, thence to deep cervical; some vessels go direct to deep cervical along superior thyroid veins; some go downwards to nodes in superior mediastinum.

Tonsillar (Waldeyer's) ring

Refers to *palatine tonsils (tonsil)*, one on each side between pillars of fauces, *lingual tonsil* on back of tongue and *pharyngeal tonsil (adenoid)* on posterior wall of nasopharynx. Protects beginning of alimentary and respiratory tracts. Consists of masses of lymphoid tissue. Adenoid drains to retropharyngeal nodes, palatine tonsil to jugulodigastric nodes and lingual to jugulo-omohyoid node.

THE LYMPHATIC DRAINAGE OF THE UPPER LIMB

The lymphatic vessels of the upper limb are arranged in a superficial and a deep set which go to the axillary nodes; a few superficial vessels join the node above the medial epicondyle.

There are two sets of lymph nodes in the upper limb—superficial and deep.

The *superficial (supratrochlear)* (1 or 2) lie superior to the medial epicondyle but may lie about middle of upper arm, medial to biceps brachii. Receive vessels from medial side of hand and forearm. Efferents to lateral axillary nodes.

The deep nodes are in the axilla (*axillary*) and receive afferents from upper limb, breast and anterior and posterior walls of trunk above level of umbilicus. They are grouped in the following way: *Lateral* medial to axillary vein receiving afferent vessels from limb; *Anterior* or *pectoral* along lateral thoracic artery; afferent vessels from breast and anterior body wall above umbilicus; *Posterior* or *subscapular* along subscapular artery receiving afferent vessels from back of trunk above umbilicus; *Infraclavicular* on clavipectoral fascia, with a few outlying nodes in deltopectoral groove, receiving afferents from radial border of limb and upper convexity of breast; *Central* deep to deep fascia in floor of axilla, draining hairy skin of axilla and receiving afferents from the other groups of nodes; *Apical* behind clavicle, in apex of axilla, receive from all other groups and drain into subclavian lymph trunk.

Lymphatic drainage of breast: Essentially into pectoral group of axillary nodes, the vessels running in pectoral fascia. In addition, from periphery of breast lymphatic vessels pass (a) upwards to infraclavicular group, (b) medially and backwards to sternal group along internal thoracic artery, (c) downwards through abdominal wall to extraperitoneal lymphatic vessels inferior to diaphragm, thence to nodes in mediastinum. Few channels via intercostal vessels to intercostal nodes. Lymphatic plexus, deep to breast in pectoral fascia, anastomoses across front of sternum with similar plexus on opposite side. Axillary tail of breast drains directly into subscapular nodes.

THE LYMPHATIC DRAINAGE OF THE ABDOMEN AND PELVIS

Lymphatic system of abdomen and pelvis may be divided into (a) lymphatic drainage of alimentary tract including liver, spleen, pancreas to pre-aortic nodes, and (b) lymphatic drainage of remaining organs and parietes to para-aortic nodes.

The system is greatly simplified by recalling that the lymphatic vessels pass from the organs to nodes along the arteries.

LYMPHATIC DRAINAGE OF ALIMENTARY TRACT

Adult retains blood supply of embryonic arrangement in which canal was slung by dorsal mesentery and its three parts were supplied as follows:

Foregut (including liver, spleen, pancreas) from coeliac trunk.

Midgut from superior mesenteric artery.

Hindgut (including upper two-thirds of anal canal) from inferior mesenteric artery.

Lymph follicles can be found in mucous membrane of all parts of alimentary canal from tonsil to anal margin (including oesophagus and stomach). Microscopic size except in

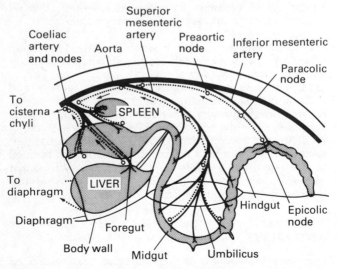

Fig. 51 Lymphatic drainage of alimentary tract.

ileum, where aggregated into visible *aggregated lymphoid follicles (Peyer's patches)* lying longitudinally in antimesenteric border of ileum. From follicles, lymphatic vessels go to nodes along alimentary attachments of mesenteries (e.g. *epicolic*), then to nodes in mesentery along arteries (e.g. *paracolic*), then to *pre-aortic nodes* related to vessels concerned (Fig. 51).

Drainage of foregut

Spleen: Lymphatics (only from capsule) go to nodes at hilum (*pancreaticosplenic*), which then drain to *retropancreatic nodes*; latter drain the pancreas. Right end of retropancreatic chain joins with *subpyloric group* (draining pylorus and head of pancreas) to end in *coeliac group.*

Liver: Into nodes in porta hepatis, except bare area, which drains through diaphragm to posterior mediastinal nodes. Lymphatic vessels from porta hepatis follow hepatic artery to coeliac nodes.

Stomach: Right half to nodes on lesser curvature to nodes along gastric arteries to coeliac nodes; left half—upper part to pancreaticosplenic and retropancreatic nodes, lower to subpyloric nodes.

Drainage of midgut

From duodenal papilla to middle of transverse colon, all drainage is via lymphoid follicles in mucous membrane to nodes along intestinal border of mesentery, then to nodes along vessels in mesentery to pre-aortic nodes at origin of superior mesenteric artery, thence to coeliac nodes.

Drainage of hindgut

From middle of transverse colon to anocutaneous junction, lymphatic vessels go from lymphoid follicles in mucosa to epicolic and paracolic nodes then to pre-aortic nodes at origin of inferior mesenteric artery; efferents thence to coeliac group. A few lymphatic vessels from lower rectum follow middle rectal artery to nodes on side wall of pelvis and along median sacral artery to nodes in hollow of sacrum.

Note: Lower one-third of anal canal drains to medial group of superficial inguinal nodes.

Coeliac nodes: Arranged round the coeliac trunk, receive all lymph from alimentary canal and drain into cisterna chyli by *intestinal lymph trunk.*

LYMPHATIC DRAINAGE OF REMAINING ABDOMINAL AND PELVIC ORGANS

Para-aortic nodes
These form (a) *medial group* alongside aorta, receiving from non-alimentary viscera and lower limb, (b) *lateral group* behind psoas, receiving from abdominal parietes. Both drain into cisterna chyli by *lumbar lymph trunk.*

External, internal and common iliac nodes
As their names suggest lie along vessels of same name. External iliac nodes receive vessels from lower limb and from most of bladder. They drain to common iliac nodes which also receive vessels from internal iliac nodes draining most of the pelvic organs. Common iliac nodes drain to para-aortic nodes.

Bladder (including prostate and lower ureter): Vessels pass with arteries to nodes along external iliac artery.

Kidneys (including upper ureter): Vessels pass to para-aortic nodes alongside 1st lumbar vertebra.

Testis: Vessels pass with arteries to para-aortic glands opposite 2nd lumbar vertebra.

Uterus and ovary: Uterus to internal iliac nodes (follow uterine artery); some vessels from body of uterus accompany vessels from ovary; some vessels from cornu may go along round ligament to superficial inguinal nodes; some vessels may pass backwards to nodes in hollow of sacrum; ovary and ampulla of tube to para-aortic nodes at origin of ovarian artery (similar to testis).

THE LYMPHATIC DRAINAGE OF THE THORAX

Parasternal nodes (6 to 10)
Along internal thoracic vessels. The afferents come from front of chest and abdominal wall, diaphragm and medial part of breast; the efferent vessels join anterior mediastinal nodes and thoracic duct.

Intercostal nodes
On heads of ribs. The afferent vessels come from chest wall; the efferent open into the thoracic and right lymphatic ducts.

Anterior mediastinal nodes (3 or 4)
Between pericardium and sternum. The afferent vessels come from lower sternal nodes, upper surface of liver and diaphragm; the efferent vessels pass to thoracic and right lymphatic ducts.

Superior mediastinal nodes (8 to 10)
In front of arch of aorta in superior mediastinum. Afferent vessels come from heart, pericardium and thymus; efferent vessels unite in *bronchomediastinal trunks* which open into the thoracic and right lymphatic ducts.

Posterior mediastinal nodes (8 to 12)
Lie along descending thoracic aorta. The afferent vessels come from the oesophagus, pericardium and diaphragm; the efferent vessels go to the thoracic duct.

Diaphragmatic nodes
Arranged on thoracic surface at periphery and divided into anterior, lateral and posterior. Receive lymphatic vessels from liver as well as diaphragm and drain to parasternal (anterior) and posterior mediastinal nodes.

Tracheobronchial nodes
Divided into *paratracheal* (sides of trachea), *tracheobronchial* (near division of trachea and between main bronchi), *bronchopulmonary* (in hilum of lung and often called hilar)

and *pulmonary* (in lung along larger bronchi). Drained by bronchomediastinal lymph trunk.

Drainage of lungs
Superficial (subpleural) and deep (bronchial) plexuses drain to tracheobronchial nodes which are drained by right and left bronchomediastinal lymph trunks, right ending in right lymph duct and left in thoracic duct.

THE LYMPHATIC DRAINAGE OF THE LOWER LIMB

The vessels of the lower limb are divided into a superficial and a deep set. The superficial, except a few which pass to the popliteal nodes, go to the superficial inguinal nodes. The deep vessels enter the deep inguinal nodes.

Popliteal nodes (4 or 5)
Are placed on the popliteal vessels. The afferent vessels come from the heel mainly, lateral side of leg and foot and knee joint to some extent; the efferents go to the deep inguinal nodes.

The superficial inguinal nodes (8 to 10)
   Oblique: Lie inferior to inguinal ligament. Lateral group receive from flank and buttock; medial group from anterior abdominal wall below umbilicus, perineum, lower one-third anal canal and vagina and external genitalia including urethra.

   Vertical or femoral: Lie along the great saphenous vein and receive the superficial vessels of the limb. The efferent vessels of both sets join the deep nodes and pass chiefly through saphenous opening.

   The deep inguinal nodes: Lie medial to femoral vein (there is usually a node in the femoral canal). The afferent vessels come from the superficial inguinal nodes, the popliteal nodes and the deep vessels of the limb; the efferent vessels join the external iliac nodes, which they reach by way of femoral canal.

THE SPLEEN

Soft, mobile, purple organ situated between left kidney and stomach in posterior upper left part of abdomen. About 12.5 cm long, 7.5 cm wide and 3.5 cm thick. Lies obliquely along line of 10th rib extending upwards to 9th and downwards to 11th rib. Lateral end lies just behind midaxillary line. Lateral (diaphragmatic) surface convex. Medial (visceral) surface has hilum in middle for entry and exit of splenic vessels. Upper part has vertical ridge dividing it into two parts, the posterior of which is related to lateral part of anterior surface of the left kidney, and the anterior to the fundus of stomach. Tail of pancreas related to visceral surface near hilum. The anterior border is notched. Inferior part of visceral surface is triangular, and rests on the phrenicocolic ligament and the left flexure of the colon.

Spleen is covered, except at the hilum, by peritoneum. Two layers of peritoneum pass forwards and medially as the gastrosplenic ligament to the fundus of the stomach. Contains left gastro-epiploic and short gastric vessels, branches of splenic vessels. Behind the hilum two layers pass on to the left kidney as the lienorenal ligament and contain the splenic vessels.

Whole of lateral surface related to diaphragm, upper two-thirds to pleura as well and upper third to lung.

Spleen has capsule which sends trabeculae into the organ. Organ consists of *white pulp* and *red pulp*. In white are ovoid masses of lymphoid tissue (appear as white dots on surface of sectioned spleen) and in the red are blood vessels (sinusoids), reticular fibres and all types of blood cells especially red blood corpuscles.

Spleen supplied by large splenic artery from coeliac trunk. Splenic vein joins superior mesenteric vein to form portal vein.

THE THYMUS

The thymus grows until puberty (about 15 g at birth to about 35 g at puberty) after which it gradually shrinks. Most of its growth takes place before the age of two years.

Thymic tissue is replaced by fat. It is situated in the superior mediastinum, and can extend downwards to level of 4th costal cartilage and upwards on the trachea as high as the lower border of thyroid gland. It consists of two lateral lobes connected by areolar tissue.

Relations

Anterior: Manubrium and sternohyoid and sternothyroid muscles.

Posterior: From above downwards, trachea, brachiocephalic artery, left brachiocephalic vein, aorta and pericardium.

May extend laterally between pleura and costal cartilages.

Thymus consists of lobules each of which has an outer cortex of lymphocytes and an inner medulla with few lymphocytes and thymic (Hassal's) corpuscles.

6
The Nervous System

This is divided into (a) *central nervous system* (spinal cord and brain), (b) *peripheral nervous system* (cranial and spinal nerves and *autonomic system—sympathetic* and *parasympathetic*).

THE CENTRAL NERVOUS SYSTEM

THE SPINAL CORD

The spinal cord is in the form of an elongated cylinder about 45 cm long lying in the vertebral canal and extending from the foramen magnum to the lower border of the 1st lumbar vertebra. Up to the third month of fetal life the cord is the same length as the canal; at birth it extends to the 3rd lumbar vertebra.

Above 2nd lumbar vertebra, cord is surrounded by membranes, including subarachnoid space and cerebrospinal fluid; spinal vessels lie within membranes; internal venous vertebral plexuses outside membranes; soft fat outside dura mater.

Below 2nd lumbar vertebra, elongated dorsal and ventral roots of spinal nerves from 2nd lumbar segment downwards contained in membranes and cerebrospinal fluid (the roots constitute the *cauda equina*); internal vertebral venous plexuses; *filum terminale.*

Note: Lumbar puncture is performed between the 3rd and 4th or 4th and 5th lumbar vertebrae because subarachnoid space extends as far as 2nd sacral vertebra.

Enlargements: The spinal cord has two enlargements at levels of brachial and lumbosacral plexuses. The *cervical* extends from the 3rd cervical to the 2nd thoracic vertebra. The *lumbar* commences opposite the 9th thoracic vertebra, is largest at the 12th thoracic, and thence tapers, forming the *conus medullaris*, from the tip of which the filum terminale descends.

Filum terminale passes from the end of the conus medullaris downwards in the middle of the cauda equina; becomes closely invested with dura mater opposite the 1st or 2nd sacral vertebra; blends with the periosteum at the lower end of the sacral canal.

Nerve roots: Formed by bundles of *ventral* and *dorsal rootlets* (4 or 5 rootlets form a root). The ventral rootlets are arranged irregularly along the side of the cord; the dorsal rootlets issue in a straight line. A *dorsal root* has a *ganglion* and is larger; the ganglia are situated in the intervertebral foramina.

On each side a ventral and dorsal root pass laterally to an intervertebral foramen in which they unite to form a *spinal nerve* (31 pairs); those in the upper part pass almost transversely; below they pass more obliquely, until in the lower part of the canal their course is vertical. The collected bundle of nerve roots at the termination of the cord form the *cauda equina.*

Spinal part of accessory nerve emerges from upper five cervical segments behind ligamentum denticulatum as a series of roots which run upwards and join together to pass through foramen magnum (p. 266).

Blood supply to spinal cord (p. 159); venous drainage (p. 195).

Anterior median fissure: Along the anterior surface of the cord in the midline. Extends into the substance of the cord for about a third of its thickness, but deeper inferiorly than superiorly. Does not reach grey matter, its floor being formed by transverse band of white matter, the *anterior white commissure.* Lined with double fold of pia mater.

Posterior median sulcus: Shallow; from it a septum of neuroglia reaches down to grey matter.

Columns of the cord: The cord is divided into two lateral halves by the median fissure and sulcus, and subdivided on each side by the nerve roots into *ventral, lateral* and *dorsal columns.* The dorsal column is subdivided by a groove (only found in the upper part of the cord), a little lateral to the posterior median fissure, into two columns, corresponding to the fasciculus gracilis (medial) and the fasciculus cuneatus (lateral).

Structure of the spinal cord (Fig. 52)

The spinal cord consists of white matter externally and grey matter internally. The grey matter is in the form of a fluted column which in transverse section appears like a letter H.

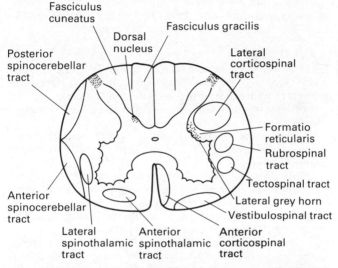

Fig. 52 T.S. thoracic spinal cord. In the white matter ascending tracts are shown on the left, descending tracts on the right.

The grey matter
The limbs of the H are crescent-shaped and are joined by the *grey commissure*.

Each crescent has two *horns*, a *ventral*, thick and short, not reaching the surface of the cord; and a *dorsal*, long and slender, reaching almost to the origin of the dorsal nerve roots where it becomes enlarged and less opaque, forming the *substantia gelatinosa*. The size of the grey crescents varies in different parts of the cord, being largest in the cervical and lumbar enlargements. In the thoracic region there is a projection of the grey matter on the lateral side of the crescent between the ventral and dorsal horns, called the *lateral horn*. It contains the cell bodies of the preganglionic sympathetic neurons (Fig. 46). *Thoracic nucleus* at base of dorsal grey horn, consists of neurons whose fibres form anterior spinocerebellar tract.

Central canal
Extends through the whole length of the cord in the middle of the grey commissure. It is lined with a ciliated epithelium. Continuous above with the 4th ventricle.

The white matter
Encloses the grey matter in each lateral half of the cord, except where the dorsal horn comes to the surface. The portion of white matter between the grey commissure and the anterior median fissure constitutes the *white commissure* (Fig. 46).

Posterior white columns (*fasciculus gracilis* and *fasciculus cuneatus*): Sensory (tactile exteroceptive and conscious proprioceptive).

Lateral white columns: Contain descending tracts (*lateral corticospinal* or pyramidal, and extrapyramidal including *rubrospinal, tectospinal, olivospinal*) and ascending (*spinocerebellar* associated with proprioception and *lateral spinothalamic*, associated with pain and temperature).

Anterior white columns: Contain descending tracts (*anterior corticospinal* and extrapyramidal including *vestibulospinal*) and ascending tracts (*anterior spinothalamic*, associated with touch).

The meninges of the spinal cord

The *dura mater* is the most external and is continuous with that investing the brain; it differs, however, from the cranial dura mater in that it comprises a single layer, is devoid of sinuses, and is separated from the vertebrae by a space which contains much fat and the internal vertebral plexuses (p. 194). It is attached above to the edge of the foramen magnum and to the membrana tectoria; from the lower border of the 2nd sacral vertebra it continues as a slender cord and blends with the periosteum of the coccyx. This dura mater gives sheaths to all the spinal nerve roots.

The *arachnoid* is placed outside the pia mater, and loosely invests the cord. Lies surface to surface with dura mater. The *subarachnoid space* of the cord is large, contains cerebrospinal fluid and is imperfectly divided by the *ligamentum denticulatum* into an anterior and a posterior portion. The posterior portion is further subdivided by the posterior median septum, which passes from the posterior sulcus backwards to the opposite part of the arachnoid. Trabeculae also pass between the nerve roots and the inner surface of the arachnoid.

The *pia mater* is less vascular, thicker and more fibrous than that investing the brain. It has an external fibrous layer of longitudinal bundles with a fold dipping into the anterior fissure. The pia mater ends in the filum terminale which lies within the prolongation of the dura mater. The *ligamentum denticulatum* is a process of pia mater, passing outwards towards the dura to which it is attached by 22 tooth-like processes situated between the origins of the spinal nerve roots; its pial origin is continuous and lies between the ventral and dorsal nerve roots.

THE BRAIN

The brain consists of the *hindbrain*, *midbrain* and *forebrain*. The hindbrain is subdivided into the *medulla oblongata, pons* and *cerebellum* and the spinal cord continues upwards as the medulla oblongata which is continuous with the deeper part of the pons. The pons becomes the midbrain. The *brain stem* refers to the medulla, pons and midbrain. The *cerebral peduncles* attach the pons to the forebrain and the cerebellum is attached to the brain stem by *cerebellar peduncles*.

The hind brain

THE MEDULLA OBLONGATA

Extends from just below the foramen magnum to the lower border of the pons on the clivus. Inferiorly it is continuous with the spinal cord, superiorly it continues into the pons, anteriorly it rests upon the clivus, and posteriorly it lies in a depression (*vallecula*) between the hemispheres of the cerebellum where it contains the lower half of the floor of the 4th ventricle.

It is about 3 cm long, 2 cm wide, 1 cm thick. It is wider near the pons. Lower half is cylindrical and closed round central canal. Upper is open and forms part of 4th ventricle (Figs 53–55).

Anterior median fissure of spinal cord continues into medulla and terminates just below the pons in the *foramen caecum*. Most of the fibres of the pyramids decussate in the lower part of the fissure, and partly interrupt it.

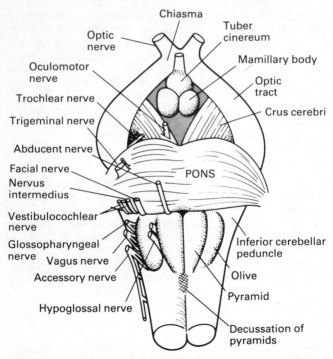

Fig. 53 Anterior view of medulla oblongata, pons and midbrain.

Structure of the medulla (Fig. 53)
If the parts of the spinal cord are traced into the medulla, its structure is more easily understood.

The line of the dorsal roots of the spinal nerves continues upwards as the line of the nerve bundles of the cranial accessory, vagus and glossopharyngeal nerves. These bundles arise from the posterolateral sulcus which if traced upwards turns outwards, so that about half-way along the medulla it is lateral, and in its upper part lies close to the posterior margin of the olive.

The part behind these nerve roots corresponds with the posterior columns (fasciculus gracilis and cuneatus) of the cord.

The line of the ventral roots of the spinal nerves, when traced into the medulla, deepens into a groove (*anterolateral*

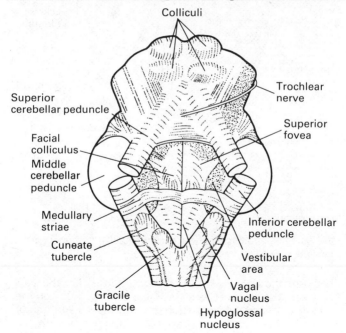

Fig. 54 Posterior view of medulla oblongata and midbrain.

sulcus) which is continued upwards nearly as far as the pons. The bundles of the nerve roots of the hypoglossal nerve issue from this groove. The part of the medulla between this groove and the anterior median fissure corresponds with the anterior column of the cord. The part

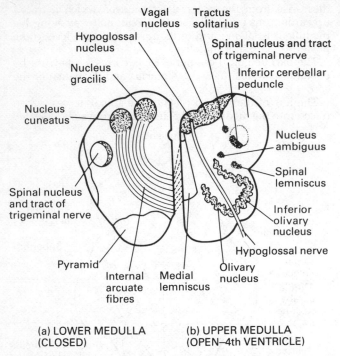

(a) LOWER MEDULLA (CLOSED)

(b) UPPER MEDULLA (OPEN–4th VENTRICLE)

Fig. 55 T.S. through (a) lower part of medulla, (b) upper part of medulla.

between the anterior and posterior regions, i.e. the part between the line of issue of the nerve bundles of the accessory, vagus and glossopharyngeal nerves and line of issue of those of the hypoglossal, corresponds with the lateral column of the cord.

Posterior part of the medulla (Fig. 54)
The posteromedial column of the cord (fasciculus gracilis) is continued into the medulla, and expands as it approaches the 4th ventricle. This expansion is called the *gracile tubercle* and on reaching the 4th ventricle the tubercles of opposite sides separate to some extent. The posterolateral column of the cord (fasciculus cuneatus) also expands as it extends upwards and forms an eminence, the *cuneate tubercle*.

Between the fasciculus cuneatus and the line of origin of the nerve bundles of the accessory, vagus and glossopharyngeal nerves, there appears in the lower part of the medulla a longitudinal prominence, which broadens out above into the *tuberculum cinereum*. This is formed by the spinal nucleus and tract of the trigeminal nerve and is continuous with the gelatinous substance of the cervical part of the spinal cord (p. 217).

Lateral part of the medulla
The tracts of the lateral column of the cord can be followed up into or down from the medulla. The lateral corticospinal tract, passes obliquely downwards from the opposite pyramid into the lateral white column of the cord. The posterior spinocerebellar tract joins the inferior cerebellar peduncle. The rest of the column (lateral spinothalamic and extrapyramidal tracts, p. 218) passes upwards and lies deep to the olive.

The *inferior cerebellar peduncle* is formed by the conjunction of a number of bundles of fibres from the spinal cord and medulla and lies lateral to the olive (Fig. 53) in the upper part of the medulla. It forms the lateral boundary of the lower part of the 4th ventricle (Fig. 54). The inferior peduncle contains the *posterior spinocerebellar tract*, *olivocerebellar* and *vestibulocerebellar fibres* and the *external arcuate fibres* from the gracile and cuneate nuclei of same and opposite sides. These fibres emerge through, medial to and lateral to the pyramid, arch over the olive, and join the inferior cerebellar peduncle. The peduncle also contains fibres from the *reticular* and *cuneate nuclei*.

Anterior part of the medulla

The *pyramids* are two elongated prominences, broader above than below. They lie on each side of the anterior median fissure. They consist of two sets of fibres, medial (four-fifths) which when traced downwards decussate (*pyramidal decussation*) and pass to the lateral white column of the spinal cord as the lateral corticospinal tract; and lateral (one-fifth) which do not cross the middle line, but descend into the anterior white columns as the anterior corticospinal tract. The crossing of the former fibres constitutes the decussation of the pyramids.

The *olive* is an oval prominence lying in the upper part of the medulla between the pyramid and the inferior cerebellar peduncle. It is separated from the pons by a groove, which contains some of the external arcuate fibres. On the medial side lie the roots of the hypoglossal nerve; and on the lateral side, but separated from it by a groove, lie the roots of the accessory, vagus and glossopharyngeal nerves.

THE FOURTH VENTRICLE

The central canal of the cord expands in the upper and posterior part of the medulla as far as the middle peduncles of the cerebellum. It gradually narrows, and becomes continuous with the cerebral aqueduct of midbrain above. The floor or anterior (ventral) wall of the ventricle is thus diamond-shaped. The lower end has been compared in shape with a pen, and is termed the *calamus scriptorius*.

The ventricle is bounded below by the diverging inferior cerebellar peduncles and above by the converging superior cerebellar peduncles. At the inferior angle are the gracile tubercles. The *roof* is tentlike and is formed mainly by pia mater inferiorly and a thin layer of white matter (*superior medullary velum*) superiorly. On the former lies the nodule and on the latter the lingula of the vermis of the cerebellum. The roof is lined by ependyma, as is the rest of the ventricle.

The floor is formed inferiorly by the upper part of the posterior surface of the medulla, and superiorly by the pons.

The floor (Fig. 54)
The floor is divided into right and left halves by a median groove.

In the lower part of each half there is a pit, the *inferior fovea*, at the side of the groove near the medullary striae. From the pit two grooves pass downwards, one towards the midline and the other towards the lateral boundary. Three areas are thus marked off in each lateral half of the medullary part of the floor.

(1) *Hypoglossal triangle:* Next to midline groove, overlying nucleus of hypoglossal nerve.

(2) *Vagal triangle:* Lateral to (1), overlying dorsal nucleus of vagus nerve.

(3) *Vestibular area:* Lateral to these, overlying nuclei of vestibular nerve. This raised area extends up into the upper half.

Separating upper part of floor from lower part:

(4) *Medullary striae:* Formerly believed auditory, are now considered external arcuate fibres. Run horizontally across middle of floor.

In upper part of floor (pons) on each side there is a pit, the *superior fovea* in line with the inferior fovea, dividing each lateral half into two areas:

(5) *Facial colliculus:* Elongated elevation medially produced by fibres of facial nerve winding over nucleus of abducent nerve (Fig. 54).

(6) *Vestibular area:* Upper part of (3). Higher up is greyblue area called *locus coeruleus.*

Deep to floor of the ventricle there are:

Trigeminal nuclei and tracts: Principal and motor nuclei in lateral part of upper half.

Descending or spinal tract of trigeminal nerve deep to floor of lower half.

Ascending *mesencephalic tract of trigeminal nerve* (proprioceptive) deep to floor of upper half.

Nucleus ambiguus lies in the medulla in the formatio reticularis deep to the lower part of floor of ventricle. It is the motor nucleus for swallowing (palate and pharynx) and vocalization (intrinsic muscles of larynx); its fibres are in glossopharyngeal, vagus and cranial accessory nerves (mainly the last).

Superior salivatory nucleus alongside facial nucleus (pons); *inferior salivatory nucleus* near dorsal vagal nucleus (medulla).

The *nucleus of the tractus solitarius* (taste to facial, glossopharyngeal and vagus nerves) also lies deep to the floor of the ventricle.

The *medial longitudinal bundle* lies on either side of the midline deep to the floor of the ventricle. It links the nuclei of the oculomotor, trochlear, abducent and accessory nerves with the vestibular nuclei and tectum.

Lateral recess
This is a lateral extension of the cavity of the ventricle at its widest part, between the cerebellum and the medulla.

The roof
This is formed by pia mater lined with ependyma. It has three apertures, one on each side in the lateral recess (*lateral aperture*), and the third at the apex of the calamus scriptorius, the *median aperture of 4th ventricle*. Through these apertures the 4th ventricle communicates with the subarachnoid space.

The *inferior medullary velum* is a downward and forward extension of the white substance of the cerebellum on the pial covering of the ventricle. Projecting into the ventricle from the roof on each side of the midline is the *choroid plexus of the 4th ventricle*.

In the upper half, the roof is formed between the converging superior peduncles of the cerebellum by the superior medullary velum which is continuous with the lingula of the cerebellum.

THE PONS

The pons lies above the medulla, inferior to cerebral peduncles and between the hemispheres of the cerebellum (Fig. 53). It is about 3 cm long and 5 cm wide.

Its ventral surface is convex and has a vertical groove in the midline for the basilar artery, transverse markings and openings for vessels. Its deeper part, smaller than the anterior, is continuous inferiorly with the deeper part of the medulla, and its central part forms the upper part of the floor of the 4th ventricle (p. 225).

Laterally transverse fibres pass outwards and backwards from it and form on each side the *middle cerebellar peduncle*.

Structure

The superficial part of the pons consists of alternating layers of transverse and longitudinal fibres, with intermingled grey matter (*nuclei pontis*). The transverse fibres are partly intercerebellar and partly fibres which come from the cerebral cortex and synapse in the nuclei of the pons before going to the cerebellum of the opposite side; the longitudinal are the pyramidal bundles which come superiorly from the cerebral peduncles and inferiorly form the pyramids. The deeper part of the pons unites the medulla with the midbrain and therefore contains ascending (spinothalamic, medial lemniscus) and descending (rubrospinal, tectospinal) tracts as well as part of the vestibular nuclei and nuclei of the trigeminal, facial and abducent nerves.

THE CEREBELLUM

The cerebellum lies in the posterior cranial fossa and is separated from the cerebrum above by the tentorium cerebelli. It consists of two lateral *hemispheres* and a median *vermis*.

The inferior surface of each hemisphere is convex, and between them there is a fossa, the *vallecula*, at the bottom of which the inferior part of the vermis lies. The medulla is in contact with the anterior part of the vallecula, and the falx cerebelli with the posterior part. The superior surface of each hemisphere is flat, and not clearly marked off from the vermis.

Lobes and fissures

The whole of the surface of the cerebellum is marked by shallow parallel sulci which demarcate the *cerebellar folia*.

Certain of the furrows are deepened to form fissures, separating cerebellar lobules. The only significant subdivision of the cerebellum is into anterosuperior and posteroinferior lobes; the fissure separating these is the *fissura prima*, which is a V-shaped fissure crossing the upper surface. The *flocculus* is a small lateral projection from the *nodule* of the vermis (*flocculonodular lobe*) and lies near the lateral foramen of the 4th ventricle, in the cerebellopontine angle. The *tonsil* is posteromedial to the flocculus and encroaches on the foramen magnum.

Structure

Grey matter: This covers the whole surface of the cerebellum and constitutes the *cortex*. In addition there are central nuclei embedded in the white matter; of these, the largest and most important is the bilateral *dentate nucleus*. Cerebellar cortex receives afferent fibres from spinal cord (associated with proprioception from muscles, joints), vestibular nuclei (associated with equilibrium) and from the cerebral cortex, fronto- and temporopontine fibres via nuclei pontis. Cerebellar cortex sends efferent fibres (axons of Purkinje cells) which synapse in dentate nucleus from which fibres pass into the superior peduncle and are associated with control of muscle tone.

White matter: This forms a central stem which branches into the cortex like a tree and also forms the cerebellar peduncles which connect the cerebellum with other parts of the brain.

Cerebellar peduncles: The inferior connects the medulla to the cerebellum; the middle connects the pons to the cerebellum; the superior connects the cerebellum to the midbrain.

Principal constituent fibres of cerebellar peduncles:

(i) *Inferior peduncle:* (1) Afferent: posterior spinocerebellar tract. (2) Afferent: vestibulocerebellar fibres (impulses from labyrinth of internal ear). (3) Afferent: external arcuate fibres from gracile and cuneate nuclei (proprioceptive). (4) Afferent: olivocerebellar fibres. (5) Afferent: reticulocerebellar fibres. (6) Auditory and visual fibres. (7) Efferent: cerebellovestibular fibres.

(ii) *Middle peduncle:* (1) Fibres from nuclei pontis of opposite side (receive fibres from cerebral cortex). (2) Commissural fibres between cerebellar hemispheres.

(iii) *Superior peduncle:* (1) Afferent: anterior spinocerebellar tract. (2) Efferent: from dentate nucleus to red nucleus of opposite side, and to thalamus.

Note: The rubrospinal fibres immediately decussate so that the cerebellum affects the same side of the body.

The midbrain (Figs. 53, 54 and 56)

This comprises the two ventral *cerebral peduncles* which connect the pons with the forebrain, and the *tectum* on the dorsal aspect. The level of the *aqueduct of the midbrain* separates the smaller dorsal tectum from the much larger peduncles.

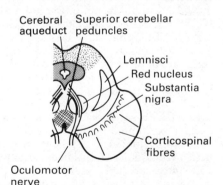

Cerebral aqueduct

Superior cerebellar peduncles

Lemnisci
Red nucleus
Substantia nigra

Corticospinal fibres

Oculomotor nerve

Fig. 56 T.S. through midbrain at level of superior colliculi.

The *cerebral peduncles* extend from the upper border of the pons and diverge into the cerebral hemispheres. Between them is the *interpeduncular fossa*, containing from behind forwards the *posterior perforated substance*, the *mamillary bodies*, *infundibulum*, and the *tuber cinereum*. Near the angle of divergence the *oculomotor nerve* issues from a groove (*oculomotor sulcus*) on the medial side of each peduncle. This groove indicates the separation of the ventral part (*crus*) from the dorsal part (*tegmentum*) of the peduncle.

There is a layer of pigmented grey matter between the basis pedunculi and the tegmentum called the *substantia nigra*.

The *crus* consists of descending fibres of which the middle third are continuous below with the pyramidal bundles of the pons and the pyramid of the medulla. Medial and lateral parts contain corticopontine fibres. All these fibres have passed through the internal capsule.

The *tegmentum* consists of ascending fibres and interspersed grey matter. Especially important ascending tracts are the medial and lateral lemnisci and the decussating superior cerebellar peduncles (Fig. 56).

Running through the length of the midbrain is the *aqueduct of the midbrain* (*cerebral aqueduct*) which connects the 3rd ventricle with the 4th. It is surrounded by grey matter, in which are the nuclei of the 3rd and 4th cranial nerves. Lateral to the aqueduct there is also the mesencephalic nucleus of the trigeminal nerve.

Third nerve nucleus lies in grey matter ventral to aqueduct at level of superior colliculi. Cranial end of this nucleus contains parasympathetic cell bodies for ciliary muscle and sphincter pupillae (p. 253). Fourth nerve nucleus lies ventral to aqueduct at level of inferior colliculi.

Level with superior colliculus, tegmentum contains *red nucleus*, perforated by emerging 3rd nerve. Scattered masses of grey matter in tegmentum constitute the *reticular nuclei*, continuous with similar formation in pons, medulla and spinal cord.

The *tectum* consists of the two pairs of *colliculi* (*quadrigeminal bodies*) situated on the dorsum of the midbrain and divided off from one another by a cruciform groove. They are reflex centres. The *superior colliculi* receive fibres from the optic tract and are a relay station between fibres in the optic tracts and cranial and spinal nerves. The *inferior colliculi* relay fibres of the lateral lemniscus (auditory pathway) to cranial and spinal nerves.

The forebrain

This is divided into a smaller part, the *diencephalon*, immediately above the midbrain consisting mainly of the thalami with the 3rd ventricle between them, and the much larger *telencephalon* consisting of the cerebral hemispheres.

THE DIENCEPHALON

The *third ventricle* is the narrow median space between the thalami, extending from the opening of the aqueduct of the midbrain inferior to the posterior commissure to the interventricular foramen. It is much deeper in front than behind.

The 3rd ventricle communicates anteriorly on each side by the *interventricular foramen* with a lateral ventricle (in a hemisphere) and behind with the 4th ventricle by the aqueduct of the midbrain.

It is roofed over by its *tela choroidea* (pia mater and ependyma) which is invaginated by a *choroid plexus* (blood vessels and ependyma), and the fornix. In the floor are the optic chiasma, tuber cinereum, infundibulum, mamillary bodies and more posteriorly the posterior perforated substance and tegmenta of the cerebral peduncles. Anteriorly is the *lamina terminalis*, the original anterior end of the neural tube. Laterally on each side is the thalamus, in front of which is the interventricular foramen bounded anteriorly by a column of the fornix. The anterior commissure is at the upper end of the lamina. Inferior to the thalamus is the hypothalamus which also extends into the floor. Posterosuperiorly are the stalk of the pineal body, the posterior commissure and the aqueduct of the midbrain.

Commissures
The *anterior commissure* is described on p. 233.

The *posterior commissure* consists of fibres connecting the two superior colliculi posteriorly. It forms the posterior boundary of the 3rd ventricle, and lies just above the opening of the aqueduct of the midbrain inferior to the pineal body.

Interthalamic adhesion or *connexus* is a delicate band of grey matter passing between the thalami, near middle of ventricle and is a secondary adhesion, not commissural.

The thalami

Two oblong masses of grey matter, covered with a thin layer of white fibres, the *stratum zonale*. Each has superior, inferior, medial and lateral surfaces and anterior and posterior extremities.

Superior surface: The lateral part of this projects into the body of the lateral ventricle; the medial part is covered by the fornix and tela choroidea. Posteriorly it is separated from the medial surface, which bounds the 3rd ventricle, by the stalk of the pineal body, and is separated laterally from the caudate nucleus by the *stria terminalis*; at the posterior end of the stria between the stalk of the pineal and the pulvinar is a triangular depressed surface, the *trigonum habenulae*.

Posterior extremity: Broad and flattened and is formed chiefly by the *pulvinar*, which overhangs the midbrain. Below and lateral to this is the eminence of the *lateral geniculate body*; more medially is the *medial geniculate body*; the *superior brachium* joins the superior colliculus with the lateral geniculate body and passes between the pulvinar and the medial geniculate body. The *inferior brachium* joins the inferior colliculus with the medial geniculate body.

Inferior surface: Lies on the tegmentum of the cerebral peduncle and receives the ascending fibres therein.

Lateral surface: Fibres of the posterior limb of the internal capsule pass between this surface and the lentiform nucleus.

Medial surface: Forms lateral wall of 3rd ventricle, and is covered by ependyma. Inferiorly is part of the hypothalamus.

Anterior extremity (anterior tubercle): Narrow and pointed; forms posterior boundary of interventricular foramen.

The *pineal body* is cone-shaped and lies at the back of the 3rd ventricle; it overlies the superior colliculi.

Note: The pineal is often calcified after middle age and then forms a useful landmark in cranial X-rays.

The *posterior perforated substance* is a depression containing grey matter in the floor of the 3rd ventricle, in the angle of divergence of the peduncles. It is bounded by the mamillary bodies in front, and by the pons behind. It is perforated for blood vessels to the thalami.

The *mamillary bodies* are two small whitish bodies in front of the posterior perforated substance and behind the tuber cinereum. Their cell bodies synapse with fibres in the fornix which come from the hippocampus, and their fibres go to the anterior nucleus of the thalamus (*mamillothalamic tract*).

The *tuber cinereum* is an eminence of grey matter, situated in front of the mamillary bodies and extending forwards to the optic tract. From its centre a tubular conical process of grey matter, the *infundibulum*, passes downwards and forwards to the posterior lobe of the hypophysis.

The *hypophysis cerebri (pituitary gland)* is a reddish-grey vascular mass lying in the sella turcica. It consists of two lobes, the anterior being the larger.

The *optic chiasma* is the union of the two optic nerves and is in the midline just in front of the tuber cinereum. From the posterior part the two optic tracts pass laterally and backwards round the cerebral peduncles.

In the chiasma optic nerve fibres from the medial half of each retina decussate; those from the lateral half remain uncrossed. Right side of the brain thus receives impulses from the right halves of the two eyes, the lateral half of the right retina and the medial half of the left, i.e. from the *left halves* of the *fields of vision*.

The *lamina terminalis* is a layer of grey matter extending from the optic chiasma inferiorly to the rostrum of the corpus callosum superiorly; laterally it is connected with the grey matter of the anterior perforated substance, which lies inferior to the anterior end of the corpus callosum, and anteriorly with the paraterminal gyrus.

The *anterior commissure* is a bundle of white fibres lying in the anterior part of the 3rd ventricle, in front of the anterior columns of the fornix. Some of the fibres (anterior) connect the two olfactory bulbs. The larger posterior part consists of fibres going through the caudate nucleus and passing under the lentiform nucleus to enter the white matter of the temporal lobe.

Note: Since the thalami are the main structures in the diencephalon, the surrounding parts are named after their relation to the thalamus. The *metathalamus* (posterior) contains the lateral and medial geniculate bodies. The *epithalamus* contains the habenula, pineal body and its stalk with the posterior and habenular commissures. The *hypothalamus* (inferior) extends from the optic chiasma to the mamillary bodies and is the most important of these.

THE CEREBRAL HEMISPHERES

The cerebral hemispheres form as a whole an oval mass, separated into approximately equal halves by the *longitudinal cerebral fissure*. Each hemisphere has three surfaces.

(1) *Superolateral:* Convex, occupying the vault of the cranium. It has a deep cleft, the lateral cerebral sulcus, between the parts in the anterior and middle fossae.

(2) *Medial:* Flat, forming one side of the longitudinal fissure.

(3) *Inferior:* Irregular, corresponding in shape to the anterior and middle fossae, and to the upper surface of the tentorium.

The surface of each hemisphere consists of grey matter and is divided into *convolutions* or *gyri* by intervening *sulci*.

The longitudinal fissure separates the two hemispheres dividing them completely in front and behind, but inferiorly near the middle they are united by the corpus callosum. The falx cerebri lies in the fissure.

The *transverse fissure of the cerebrum* is an infolding of the cortex of the brain and is seen when the tela choroidea and choroid plexuses of the lateral ventricles are removed. It extends from the tip of the inferior horn of the lateral ventricle on one side, over the thalami and 3rd ventricle to the tip of the inferior horn on the other side.

The *choroid fissure* is C-shaped, extending from the tip of the inferior horn round the thalamus to the interventricular foramen. It accommodates the choroid plexus of the lateral ventricle. The two choroid fissures together constitute the transverse fissure of the cerebrum.

SOME OF THE MAIN SULCI

(1) The *lateral sulcus* commences at the lateral side of the anterior perforated substance, passes outwards through the whole depth of the hemisphere and divides into three branches, short *anterior* and *ascending*, both passing into the frontal lobe, and a long *posterior* passing upwards and backwards to about the middle of the superolateral surface of the hemisphere. On separating the edges of this sulcus the *insula* is exposed.

(2) The *central sulcus* commences at the longitudinal fissure near the vertex and passes downwards and forwards nearly as far as the division of the lateral sulcus.

(3) The *parieto-occipital sulcus* lies on the medial and superolateral surfaces of the brain. The medial part is perpendicular and about one-third anterior to the posterior end of the hemisphere and continues into the lateral part for about 3 cm.

(4) The *calcarine sulcus* is on the posterior part of the medial surface extending from the posterior pole forwards to meet the parieto-occipital sulcus and continue as far as the posterior end of the corpus callosum.

(5) The *sulcus cinguli* commences in front near the anterior perforated substance, takes a course on the medial surface of the hemisphere about midway between the corpus callosum and the superior edge of the hemisphere and ends a little behind the upper end of the central sulcus.

(6) The *collateral sulcus* lies on the inferior surface parallel to and about 2 cm from its medial edge; it extends into the hemisphere and forms the *collateral eminence* in the inferior horn of the lateral ventricle.

(7) *Circular sulcus* delineates the *insula*, a buried area of grey matter (cortex) in the depths of the lateral sulcus (p. 238).

THE LOBES

The hemispheres are divided into the following lobes:

(1) *Frontal:* Limited below by the lateral sulcus, behind by the central sulcus, medially by sulcus cinguli.

(2) *Parietal:* Limited in front by the central sulcus, and behind by the parieto-occipital sulcus extended downwards to the *pre-occipital notch* about 3 cm from the occipital pole; below by lateral sulcus continued backwards, and medially by sulcus cinguli.

(3) *Occipital:* Bounded anteriorly by the parieto-occipital sulcus; forms the posterior part of the hemisphere on all surfaces.

(4) *Temporal:* Occupies middle fossa at the base of the skull and lies below lateral sulcus and in front of line joining parieto-occipital sulcus to pre-occipital notch.

(5) *Insular (insula):* Buried within lateral sulcus.

(1) *Frontal lobe*
Inferior surface: Divided by the *orbital sulcus* (H-shaped) into medial, lateral, anterior and posterior *orbital gyri*. On the medial orbital gyrus is the *olfactory sulcus*, which contains the olfactory bulb and tract. Medial to this is the *gyrus rectus*.

The *frontal gyri* are three in number—*superior, middle* and *inferior*. They lie horizontally on this surface separated from each other by two sulci, and lie in front of the precentral sulcus.

The inferior frontal gyrus is divided by the branches of the lateral sulcus into *pars orbitalis* (inferior), *pars triangularis* (middle) and *pars opercularis* (posterior).

Medial surface: The *medial frontal gyrus* extends along the edge of the longitudinal fissure. It is limited below by the sulcus cinguli, and behind by a short vertical sulcus which separates it from the *paracentral lobule*, bounded behind by the upturned end of the sulcus cinguli. Below this the *gyrus cinguli* extends down to the *callosal sulcus* and forms the lower part of the medial surface of the frontal and parietal lobes.

(2) *Parietal lobe*
Superolateral surface: The *intraparietal sulcus* commences

near posterior limb of lateral sulcus, passes upwards parallel to the lower half of the central sulcus as *inferior postcentral sulcus*, and then turns backwards to end near parietooccipital sulcus.

The *superior postcentral sulcus* continues the line of the vertical part of the intraparietal sulcus.

The *postcentral gyrus* is bounded in front by the central sulcus, behind by the *postcentral sulcus*, below by the lateral sulcus.

The *superior parietal lobule* is bounded by the postcentral sulcus in front, behind by the parieto-occipital sulcus. The *inferior parietal lobule* is subdivided into an anterior part (*supramarginal gyrus*) bounded in front and above by the intraparietal sulcus and behind by the posterior limb of the lateral sulcus, a middle part (*angular gyrus*) in the centre of the superolateral surface of the parietal lobe (below the superior parietal lobule, above the temporal lobe and behind the occipital lobe), and a posterior part curving over the upturned middle temporal sulcus, and continuing into the inferior temporal gyrus.

Medial surface: The *precuneus* lies anterior to the *cuneus* and is bounded behind by the parieto-occipital sulcus, and in front by the ascending terminal limb of the sulcus cinguli.

Inferior surface: Above the lateral sulcus; with part of frontal lobe forms *frontoparietal operculum.*

(3) *Occipital lobe*

Superolateral surface: Three *gyri, superior, middle* and *inferior occipital*, separated by two sulci—*transverse* and *lateral occipital.*

Medial surface: The *cuneus* is the area between the parietooccipital and the calcarine sulci.

The *calcarine sulcus* commences at the posterior part of the medial surface of the occipital lobe by a forked extremity; is joined about half-way by the parieto-occipital sulcus, and ends near the posterior extremity of the corpus callosum. Below this is the *lingual gyrus* continuous in front with the *hippocampal gyrus.*

Inferior surface: The *medial* and *lateral occipitotemporal gyri* extend from the pole of the temporal lobe to the posterior part of the hemisphere forming the inferior surface of both temporal and occipital lobes. The medial gyrus is separated by the *collateral sulcus* from the lingual gyrus which completes this surface.

(4) *Temporal lobe*
Superolateral surface: Lies in contact with the insula forming the *temporal operculum.*

The *superior* and *inferior temporal sulci* run parallel to posterior limb of the lateral sulcus. The *superior temporal gyrus* is bounded above by posterior limb of lateral sulcus, and is continuous behind with anterior part of the inferior parietal lobule. The *middle temporal gyrus* lies between the temporal sulci and joins the middle part of the inferior parietal lobule; behind it merges into the middle occipital gyrus.

The *inferior temporal gyrus* (below inferior temporal sulcus) joins the inferior occipital gyrus behind.

Inferior surface: Occipitotemporal gyri, described with occipital lobe.

(5) *Insular lobe (insula)*
A submerged triangular area of cortex only exposed by separating the gyri bounding the lateral sulcus. These gyri form the *opercula—orbital* (pars orbitalis), *frontal* (pars triangularis), *frontoparietal, temporal* formed by upper surface of temporal lobe. The island is limited anteriorly, superiorly and postero-inferiorly from the overlying gyri by the *circular sulcus,* but antero-inferiorly at the *limen insulae* its gyri are in continuity with the anterior perforated substance. The *sulcus centralis insulae* running upwards and backwards from the limen separates the *short gyri* (superior) from the *long gyrus* (inferior).

THE RHINENCEPHALON

This includes those parts of the brain concerned with olfaction (also called the *smell-brain*). In man many parts of the rhinencephalon have functions other than smell, such as emotions and memory. The term *limbic system* is used to include a large number of structures as well as those associated with smell. The *olfactory bulb* and *tract* lie in the olfactory sulcus on the orbital surface of the frontal lobe. Posteriorly the tract bifurcates. The *lateral stria* passes across the commencement of the lateral sulcus to the *uncus* of the *hippocampal gyrus*, and the *medial* to the longitudinal fissure to join the *medial* and *lateral longitudinal striae* on the *indusium griseum*.

The indusium griseum, a thin sheet of grey matter lying on upper surface of corpus callosum, is continuous with *gyrus cinguli* round the *callosal sulcus* which is immediately above corpus callosum. The indusium griseum commences near the anterior perforated substance as a continuation of the medial stria of the olfactory tract. It follows the curve of the corpus callosum near the posterior end of which it becomes continuous with the *dentate gyrus*.

The *dentate gyrus* lies in the *hippocampal sulcus*. It is formed by the superficial grey matter of the hemisphere which here ends in a fringed margin.

The *parahippocampal (hippocampal) gyrus* commences below the splenium of the corpus callosum; here the end of the calcarine sulcus cuts into it, leaving a narrow isthmus connecting it with the indusium griseum. It runs forwards above the collateral sulcus which separates it from the inferior surface of the temporal lobe and ends as the uncus.

The *hippocampal sulcus* lies along the upper border of the hippocampal gyrus, and projects into the inferior horn of the lateral ventricle, forming the *hippocampus*.

Structures included in the limbic system are the *septum pellucidum* (p. 243), *amygdaloid body* (p. 245), *stria terminalis* (p. 245), *anterior perforated substance, hippocampus, fornix* (p. 243), *mamillary bodies, anterior part of thalamus, gyrus cinguli*, a large part of the *hippocampal gyrus* and probably the *habenular structures* near the pineal stalk.

THE INTERIOR OF THE CEREBRUM

Corpus callosum
Lies inferior to the longitudinal fissure. It is the large commissure of the hemispheres and consists principally of transverse fibres. It is about 10 cm long and extends to within 4 cm of the anterior and 6 cm of the posterior end of the hemispheres.

It is broader behind than in front and thicker at each end than in the middle. In sagittal section the corpus callosum is thickened posteriorly to form the *splenium* and bent downwards anteriorly to form a *genu*. Between the splenium and genu is the *body*. From the genu a thin layer passes downwards and backwards towards the anterior commissure.

On either side the fibres of the corpus callosum radiate into the hemisphere (*radiation of the corpus callosum*). Those passing forwards into the frontal lobes are called the *forceps minor* and those forming the splenium and passing into the occipital lobes, the *forceps major*. The fibres on the superolateral aspect of the occipital (posterior) horn of the lateral ventricle are known as the *tapetum*.

Its upper surface is convex and covered by the indusium griseum, on which are the delicate ridges formed by the medial and lateral longitudinal striae.

Relations: Upper surface forms floor of longitudinal fissure; inferior surface connected behind with the fornix, and in front of this with the septum pellucidum. The extremity of the rostrum is connected inferiorly with the lamina terminalis. The corpus callosum forms the roof of the bodies and anterior horns of the lateral ventricles.

Lateral ventricles
One in each hemisphere. Each communicates anteriorly with the third ventricle by an interventricular foramen and consists of a central part, the *body*, and three *horns*, *anterior*, *posterior* and *inferior* (Figs 57 and 58).

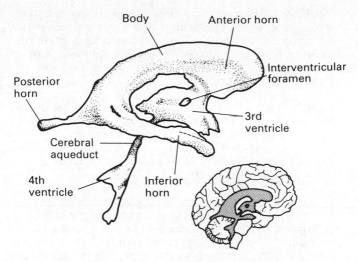

Fig. 57 Ventricles of brain viewed from right.

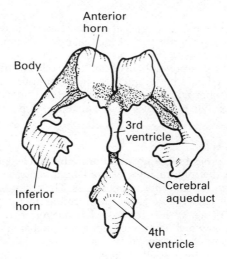

Fig. 58 Ventricles of brain viewed from in front.

Relations of the body: Roof, tapetum of corpus callosum; medial wall, septum pellucidum and fornix (anterior part); floor from before backwards: (1) caudate nucleus of corpus striatum, (2) stria terminalis, (3) part of thalamus in front of choroid plexus, (4) choroid plexus of lateral ventricle, (5) fimbriated edge of the fornix.

The *anterior horn* curves forwards and somewhat lateral and downwards into the frontal lobe over the caudate nucleus, anterior to the interventricular foramen.
Relations: Anterior—genu of corpus callosum; roof—corpus callosum; inferior and lateral—caudate nucleus; medial—septum pellucidum.

The *posterior horn* projects backwards and medially into the occipital lobe. Usually asymmetrical, may be absent.
Relations: Superior and lateral—tapetum of corpus callosum; inferior—white substance of occipital lobe; medial—*calcar avis* (due to inward projection of calcarine sulcus); at junction of posterior and inferior horns—*collateral trigone* (due to collateral sulcus).

The *inferior horn* passes at first backwards and outwards round the posterior part of the thalamus, then downwards and forwards into temporal lobe.
Relations: Superior—tail of caudate nucleus, stria terminalis, amygdaloid body; inferior—hippocampus with alveus, fimbria, choroid plexus, collateral trigone; medial—pia mater, ependyma; lateral—white substance of temporal lobe.

Hippocampus
A large projection of grey matter in the whole length of the floor of the inferior horn of the lateral ventricle. It corresponds with the hippocampal sulcus. The anterior extremity becomes enlarged and grooved, forming the *pes hippocampi*. Its ventricular surface is covered by a layer of white matter, the *alveus*, which continues along the medial border of the hippocampus as the *fimbria*, a narrow white band prolonged into the *crus of the fornix*.

Fornix
Consists of two longitudinal bundles of fibres inferior to the
corpus callosum. The bundles are separated in front forming the *columns*, and joined in the middle, forming the *body*.

The body is triangular in shape, the base being placed
posteriorly where it is close to the corpus callosum. In front
of this its upper surface is connected with the septum
pellucidum. The lateral margins are free, lying above the
choroid plexus of the lateral ventricles. The inferior surface
lies upon the tela choroidea of 3rd ventricle.

The columns pass downwards forming the anterior
boundary of the interventricular foramen, then through the
grey matter on the lateral side of the 3rd ventricle to end in
the mamillary bodies where they synapse.

The *mamillothalamic tract* goes to the anterior nucleus of
the thalamus. Some of the fibres in the columns are
connected with the anterior perforated substance and
hypothalamus.

The fimbria of each side is prolonged upwards along
medial ependymal wall of inferior horn. Curving around
pulvinar it becomes the *crus of fornix* which is attached to
splenium of corpus callosum. The two crura converge to
the midline and form body of fornix. Most fibres proceed
through body to columns of fornix, but a few cross midline,
the *commissure of the fornix*, to reach opposite mamillary
body.

Interventricular foramen
An opening between the column of the fornix and the
thalamus. It is the communication between 3rd and lateral
ventricles.

Septum pellucidum
A double vertical partition between the lateral ventricles,
attached to the fornix below and to the inferior surface of
the body of the corpus callosum above. Between the layers
is the *cavity of septum pellucidum*, a closed narrow space not
connected with the ventricles.

Tela choroidea of the 3rd ventricle
Triangular double fold of pia mater prolonged through the choroid fissure and lying over the 3rd ventricle and upper surfaces of the thalami. Its apex reaches the interventricular foramen, and the fornix lies upon its upper surface. Between its layers are:

(1) *Choroid plexuses:*
Of the lateral ventricles: Fringed vascular processes extending from the interventricular foramen to commencement of inferior horn, and attached along the lateral margins of the tela choroidea of this ventricle; thence each passes into the inferior horn, resting on the fimbria and hippocampus, forming the *choroid plexus of the inferior horn.* Supplied by anterior choroidal artery from internal carotid, reinforced by posterior choroidal branches from posterior cerebral.

Of the 3rd ventricle: From the under surface of the tela choroidea there depend two vascular fringes, diverging behind and forming the choroid plexuses of 3rd ventricle.

(2) *Internal cerebral veins:* Two veins each formed by junction of the *thalamostriate vein*, veins of *septum pellucidum* and *choroidal vein* of each side; run backwards between the layers of the tela choroidea and unite posteriorly to form the *great cerebral vein*, which opens into the straight sinus.

THE BASAL GANGLIA

These refer to masses of grey matter anterior and lateral to the thalamus within each hemisphere. The thalamus is not usually included in this term. They are (a) *caudate nucleus*, (b) *lentiform nucleus*, (c) *claustrum* and (d) *amygdaloid body*. (a) and (b) form the *corpus striatum*, so called from its appearance on section which displays alternate white and grey bands. The larger, the lentiform nucleus, lies in the white matter of the hemisphere deep to the insula, and the smaller caudate nucleus lies in the floor of the lateral ventricle anterior and medial to the lentiform.

The *caudate nucleus* is comma-shaped, with the larger end (*head*) anterior in the floor and lateral wall of the anterior horn; the narrow posterior part (*body* and *tail*) lies along the floor and lateral wall of the lateral ventricle and makes a

C-shaped curve into the roof of the inferior horn; it is accompanied by the stria terminalis which joins the amygdaloid body in the roof of the inferior horn.

The *lentiform nucleus* lies lateral to and at a lower level than the caudate nucleus from which it is separated by the anterior limb of the internal capsule which consists of ascending and descending fibres. The lentiform nucleus consists of a small medial part that is pale, the *globus pallidus*, and a larger lateral darker part called the *putamen*. Lateral to the nucleus is another stratum of white fibres, the *external capsule*, and lateral to this a thin lamina of grey matter called the *claustrum*, which lies deep to the insula. The lentiform nucleus is separated by the posterior limb of the internal capsule from the thalamus which lies posteromedially.

The *stria terminalis* is a narrow band of white fibres attached to the anterior column of the fornix in front, whence it passes backwards in the floor of the lateral ventricle, between the caudate nucleus and the thalamus; it then passes into the roof of the inferior horn, at the end of which it enters a mass of grey matter, the *amygdaloid body* which is continuous with the superficial grey matter at the anterior perforated substance.

THE WHITE MATTER OF THE CEREBRAL HEMISPHERES

Deep to the grey matter of the cerebral cortex most of the hemisphere consists of white matter. These fibres are grouped according to their course and connections—commissural, association, and projection fibres.

Commisural fibres
Connect corresponding areas in the two cerebral hemispheres. These are the (a) corpus callosum (p. 240), (b) anterior commissure (p. 233), (c) commissure of the fornix (p. 243), and (d) posterior commissure (p. 231).

Association fibres
Connect gyri of the same hemisphere; may be adjacent or more widely separated and are divided into *short* and *long*.

Examples of long association fibres are the *cingulum* in the gyrus cinguli from frontal lobe sweeping round corpus callosum and down into temporal lobe, the *uncinate bundle* from frontal to temporal and occipital lobes, the *superior* (between frontal and occipital lobes) and *inferior* (between temporal and occipital lobes) *longitudinal bundles.*

Projection fibres
Connect the cerebral cortex with the lower parts of the brain and the spinal cord. The projection fibres of the more primitive parts of the brain (*archipallium*) form two tracts, the fornix and the stria terminalis. The projection fibres to and from the cortex form the *corona radiata* and converge on the corpus striatum and thalamus between which they pass and form the *internal capsule.*

The *internal capsule* in a horizontal section through the hemisphere at the level of the insula appears V-shaped with an *anterior limb* between the lentiform and caudate nuclei and a *posterior limb* between the lentiform nucleus and the thalamus. The apex of the V is called the *genu.*
  The anterior limb of the internal capsule contains (a) frontopontine fibres, and (b) anterior thalamic radiation.
  The genu and posterior limb contain (a) corticospinal fibres (motor), (b) thalamocortical fibres (general sensory), (c) temporopontine fibres, (d) fibres of the auditory radiation, (e) fibres of the optic radiation, and (f) central thalamic radiation.
  (c), (d) and (e) are sublenticular rather than thalamolenticular.
  Note: The fibres in the genu end in relation to nuclei of motor cranial nerves and are called *corticonuclear*. Adjacent to these are the *corticospinal* fibres associated with the upper limb next to which are the fibres for the lower limb. This is a common site of cerebral haemorrhage.

THE MEMBRANES OR MENINGES
The dura mater
The most external; a dense fibrous membrane closely attached to the bones of the skull, forming their internal

periosteum. The inner surface is smooth and covered with mesothelium. It is continuous with the dura mater of the spinal cord through the foramen magnum. The fibrous part of the dura mater is divided into two layers; an outer, forming the periosteum, and an inner, outside the mesothelium. The inner forms the *falx cerebri* and *tentorium cerebelli* and, by its separation in certain situations, the venous sinuses (p. 185). On the upper surface, near and projecting into the superior sagittal sinus, are the *arachnoid granulations*, which are enlarged *villi of the arachnoid* projecting through the layers of dura mater.

The *falx cerebri* is sickle-shaped and lies vertically between the cerebral hemispheres; narrow end in front attached to crista galli and posterior part to the upper surface of the tentorium; superior border attached to midline of internal surface of skull; lower border is concave and free anteriorly. Between its layers in upper border is the superior sagittal sinus. In inferior free border is inferior sagittal sinus. Straight sinus is in edge attached to tentorium.

The *tentorium cerebelli* is a tent-like fold of dura mater between the cerebral and the cerebellar hemispheres. It has an outer convex border attached in front on each side to the posterior clinoid process and the superior edge of the petrous bone, and behind to the margins of the groove for the transverse sinus. The inner concave border is free posteriorly, and forms the opening through which the cerebral peduncles and the posterior cerebral arteries pass from the posterior into the middle cranial fossa; in front this border passes over the attached border to the anterior clinoid processes. The tentorium is attached to the falx cerebri; it is highest in front, and from this point descends on all sides. Between its layers are the transverse sinus from the internal occipital protuberance to the petrous temporal, superior petrosal sinus on the upper border of the petrous temporal and straight sinus in junction with falx.

The *falx cerebelli* extends vertically from the tentorium to the foramen magnum and separates the two cerebellar hemispheres. It is attached posteriorly to the internal occipital crest, where it encloses the occipital sinus, and below to each side of the foramen magnum.

The pia mater

Consists of a fibrous membrane supporting blood vessels, closely invests the brain and dips into the sulci. At the transverse fissure it is prolonged into the lateral ventricles and over the 3rd ventricle, pushing the ependymal lining of those cavities in front of it, and forming the tela choroidea carrying the choroid plexuses of the lateral and 3rd ventricles. It is prolonged over the roof of the 4th ventricle, sending inwards two vascular fringes, the choroid plexuses of that cavity.

The arachnoid (mater)

A thin membrane lying outside the pia mater in contact with inner surface of dura. Between the pia mater and the arachnoid is the *subarachnoid space*, containing the cerebrospinal fluid. The potential space between the dura mater and arachnoid is known as the *subdural space*.

The subarachnoid space is larger in some places than in others. The arachnoid stretches across between the two temporal lobes at the base of the brain, forming the *interpeduncular cistern*, which lies anterior to the pons, reaches as far forward as the optic nerves and contains the cerebral arterial circle. Between the inferior surface of the cerebellum and the posterior surface of the medulla it forms the *cerebellomedullary cistern*. This cistern can be tapped by passing a needle between the occipital bone and the atlas through the posterior atlanto-occipital membrane and spinal dura mater.

The cerebral ventricles communicate with the subarachnoid space by the median aperture of 4th ventricle, an opening in the pia mater across the roof of the 4th ventricle. There are two lateral apertures in the pia mater of this ventricle on each side near the upper roots of the glossopharyngeal nerve deep to the flocculus.

Cerebrospinal fluid is produced by the choroid plexuses of the ventricles, mainly those of the lateral. The fluid passes through the interventricular foramen into the 3rd ventricle which has its own choroid plexus. The fluid then passes through the cerebral aqueduct into the 4th ventricle and central canal of the spinal cord. The 4th ventricle has its own choroid plexus. Cerebrospinal fluid leaves the system by the foramina in the roof of the 4th ventricle and passes into the subarachnoid space. Most of it is eventually absorbed into the venous system via arachnoid villi and granulations.

Fluid is also absorbed in lymph vessels and veins surrounding cranial and spinal nerves round which there are extensions of the subarachnoid space, for example, along the optic nerve.

7
The Peripheral Nerves

THE CRANIAL NERVES

OLFACTORY (1st CRANIAL) NERVE

Nerve fibres arise from nerve cells in the nasal mucosa (roof and adjacent parts of lateral wall and septum) and form bundles which pass upwards through foramina in the cribriform plate of the ethmoid to the olfactory bulb where they synapse with the cell bodies of neurons whose fibres pass backwards as the olfactory tract (p. 239).

OPTIC (2nd CRANIAL) NERVE

Fibres from ganglion cells in retina pass through lamina cribrosa of sclera to form optic nerve. This passes backwards to the optic canal. Its fibres are myelinated but have no endoneurium. Optic nerve is really a projection of white matter of cerebrum. It is surrounded as far as the eyeball by dura mater, arachnoid, pia mater and cerebrospinal fluid, in continuity with that of the cerebral subarachnoid space. In the cranium the two nerves partially decussate to form the optic chiasma (p. 233) from which the optic tracts continue backwards to the lateral geniculate body and superior colliculus.

OCULOMOTOR (3rd CRANIAL) NERVE

Emerges from medial side of cerebral peduncle just in front of pons. Fibres come from nucleus in ventral part of grey matter round aqueduct at level of superior colliculus. Parasympathetic neurons (*Edinger-Westphal nucleus*) at cranial end of main nucleus.

Enters cavernous sinus between anterior and posterior clinoid processes and runs in lateral wall above trochlear nerve. As it passes forwards to enter orbit through superior orbital fissure, the trochlear and frontal nerves become superior to it. It divides in the fissure into two branches which enter orbit between the heads of the lateral rectus within fibrous ring; the nasociliary branch of the trigeminal lies between the two divisions.

Superior branch supplies superior rectus and levator palpebrae superioris. Inferior divides into three, for medial rectus, inferior rectus and inferior oblique. The latter gives off a branch to the ciliary ganglion where it synapses. Postganglionic fibres from ganglion supply the ciliary muscle and the sphincter pupillae.

TROCHLEAR (4th CRANIAL) NERVE

Emerges from superior medullary velum in which it decussates, and winds round lateral side of cerebral peduncle to ventral surface of midbrain. Fibres come from nucleus in ventral part of grey matter round aqueduct at level of inferior colliculus.

Enters cavernous sinus with oculomotor nerve, passes forwards in lateral wall of cavernous sinus below oculomotor; enters orbit through superior orbital fissure lateral to the tendinous ring. Passes medially over levator palpebrae superioris to supply superior oblique.

TRIGEMINAL (5th CRANIAL) NERVE

Emerges from lateral part of pons by small motor (medial) and large sensory roots. Sensory fibres end in (a) *main sensory nucleus* in dorsilateral part of pons, (b) *descending or spinal nucleus* extending downwards from main nucleus through pons and medulla to upper cervical segments, (c) *mesencephalic nucleus*, side of aqueduct of the midbrain at level of inferior colliculus (cell bodies are those of 1st neurons). *Motor nucleus* is medial to main sensory nucleus.

The two roots pass forwards below tentorium through oval opening in dura mater near apex of petrous temporal. Dura prolonged round roots and ganglion between fibrous and endosteal layers of middle cranial fossa, forming *cavum trigeminale.* Sensory root enters the trigeminal ganglion lodged in front of apex of petrous temporal. The motor root passes under ganglion and goes to foramen ovale, where it unites with the mandibular division.

Trigeminal ganglion
Lodged in a depression in front of apex of petrous temporal postero-inferior to cavernous sinus; gives off from its anterior edge the *ophthalmic, maxillary* and *mandibular divisions.* The two former are entirely sensory; the last is a mixed nerve.

There are many communicating branches between the ophthalmic nerve branches and the motor nerves in the orbit and between the mandibular nerve branches and the facial and hypoglossal nerves.

Ophthalmic nerve

Passes along lateral wall of cavernous sinus below oculomotor and trochlear nerves, to enter orbit through superior orbital fissure. In sinus communicates with carotid sympathetic plexus and oculomotor, trochlear and abducent nerves, and divides into three branches, two of which, *frontal* and *lacrimal*, enter orbit above lateral rectus outside fibrous ring, whilst the third (*nasociliary*) passes between two heads of origin of that muscle, and between two divisions of oculomotor nerve, inside fibrous ring.

Lacrimal
Passes along lateral wall of orbit to inferior surface of lacrimal gland; communicates with zygomatic of maxillary; receives parasympathetic secretomotor fibres from the pterygopalatine ganglion for lacrimal gland; finally pierces orbital fascia, and supplies skin of upper eyelid and communicates with branches of facial.

Frontal
Largest branch, enters orbit just lateral to trochlear nerve, passes forwards on levator palpebrae superioris and di-

vides into *supratrochlear*, which is directed forwards and medially between trochlea and supra-orbital notch and is distributed to skin of forehead and communicates with infratrochlear, and *supra-orbital*, a continuation of frontal, passing through notch and supplying palpebral filaments; ends on forehead by dividing just outside orbit into two branches which supply the anterior half of the scalp.

Nasociliary

Enters orbit between the heads of the lateral rectus and passes forwards and medially over optic nerve to medial wall of orbit. Continues as *anterior ethmoidal nerve* through anterior ethmoidal foramen; re-enters cranium and passes downwards by side of crista galli; divides into terminal branches, medial, supplying mucous membrane of the adjacent part of the septum, and lateral, running along groove on internal surface of nasal bone and supplying lateral wall of nose. This branch passes between bone and lateral cartilage to supply skin of ala and tip of nose (*external nasal nerve*).

In the orbit the nasociliary nerve gives off:

(1) *Ganglionic branch:* Passes along lateral side of optic nerve and enters the ciliary ganglion forming its sensory root.

(2) *Long ciliary nerves:* Pass along medial side of optic nerve, join some short ciliary branches from ganglion, and, piercing the sclera, are distributed to ciliary body and iris. They also carry sympathetic fibres from carotid plexus to dilatator pupillae; cell bodies of sympathetic fibres are in superior cervical ganglion.

(3) *Posterior ethmoidal nerve:* Enters posterior ethmoidal foramen and supplies posterior ethmoidal and sphenoidal sinuses.

(4) *Infratrochlear nerve:* Passes forwards on medial wall of orbit and supplies lacrimal sac, skin of upper eyelid and root of nose.

Ciliary ganglion

Small reddish-coloured body lateral to optic nerve at the back part of the orbit.

Nerves entering ganglion are:

(1) *Sensory root:* From the nasociliary with cell bodies in trigeminal ganglion; sensory to eyeball.

(2) *Parasympathetic root:* From inferior branch of the oculomotor nerve; cell bodies in anterior part of third nerve nucleus (Edinger-Westphal nucleus); synapse in the ganglion and postganglionic fibres supply sphincter pupillae and ciliary muscle.

(3) *Sympathetic:* From the carotid plexus; frequently enter with the sensory root; cell bodies in superior cervical ganglion; fibres are vasoconstrictor to intrinsic vessels of eyeball.

The nerves leaving the ganglion are called the *short ciliary nerves*, ten or twelve in two bundles, large inferior and small superior. Pass forwards above and below optic nerve, with long ciliary nerves of nasociliary. The nerves separate and pierce the sclera round the optic nerve; run in grooves on its internal surface to end in ciliary muscle and sphincter pupillae (parasympathetic) and cornea (sensory).

Maxillary nerve

Leaves middle of trigeminal ganglion and passes through foramen rotundum into pterygopalatine fossa; enters orbit through inferior orbital fissure, runs along floor of orbit and enters the infra-orbital canal as the *infra-orbital nerve*. Emerges on the face from the infra-orbital foramen under the levator labii superioris and divides into a number of branches to supply skin of lower eyelid, medial part of cheek, lateral edge of naris, and upper lip, *nasal, inferior palpebral* and *superior labial*. Joins with branches of the facial nerve to form the infra-orbital plexus.

In the skull maxillary nerve gives off a *meningeal branch.*

In the pterygopalatine fossa it gives off:

(1) *Zygomatic branch:* Enters orbit by inferior orbital fissure and divides into (a) *zygomaticotemporal* which passes along lateral wall of orbit, enters a foramen in the zygomatic bone, and becomes cutaneous on the temple (may carry postganglionic parasympathetic fibres from pterygopalatine ganglion to lacrimal gland), and (b) *zygomaticofacial*

which passes to lower and lateral angle of orbit, goes through zygomatic foramen and becomes cutaneous on the face.

(2) *Ganglionic branches:* To pterygopalatine (sphenopalatine) ganglion.

(3) *Posterior superior alveolar (dental):* Comes off as maxillary nerve enters orbit and supplies the molar teeth and related gum and lining of maxillary sinus.

The *infra-orbital nerve* gives off *middle* and *anterior superior alveolar (dental)* which descend in canals in wall of sinus and supply the premolar teeth (middle nerve not always present) and incisor and canine teeth and anterior part of inferior concha (anterior nerve).

Pterygopalatine (sphenopalatine) ganglion
Deeply placed in pterygopalatine fossa, near the sphenopalatine foramen.

Nerves entering ganglion are:

(1) *Parasympathetic:* Preganglionic cell bodies in superior salivatory nucleus, fibres in nervus intermedius of facial; leave facial (geniculate) ganglion in middle ear as *greater petrosal nerve.* In foramen lacerum joined by *deep petrosal nerve* (sympathetic from carotid plexus) to form *nerve of pterygoid canal* which joins the ganglion. These parasympathetic fibres synapse in ganglion; postganglionic fibres are secretomotor to lacrimal gland and all glands of nose, palate and paranasal sinuses (*ganglion of hay fever* characterized by tears and running nose).

(2) *Sympathetic:* Cell bodies in superior cervical ganglion, fibres in carotid plexus forming deep petrosal nerve in foramen lacerum and joining greater petrosal. Reach ganglion via nerve of pterygoid canal and pass through ganglion without synapsing. Vasoconstrictor to mucous membrane of nose, palate, paranasal sinuses.

(3) *Sensory:* From maxillary with cell bodies in trigeminal ganglion.

The branch from the maxillary nerve leaves the ganglion as:

(1) *Greater palatine nerve* which passes downwards through greater palatine canal to hard palate where it

divides into branches which run forward in grooves in the
bone almost as far as the incisor teeth. It joins the *nasopalatine (long sphenopalatine) nerve* in the region of incisive canal
and supplies the gums and mucous membrane of hard
palate. While in the greater palatine canal it gives off *inferior
nasal branches* which supply mucous membrane on middle
and inferior nasal conchae and lining of maxillary sinus.

(2) *Lesser palatine nerves* which enter palatine canal and
supply mucous membrane of soft palate, uvula and tonsil.

(3) *Posterior superior lateral nasal (short sphenopalatine)
nerves* which pass through sphenopalatine foramen into
nasal cavity and supply the posterior and upper part of the
lateral wall of the nose.

(4) *Nasopalatine (long sphenopalatine)* which is larger and
longer than the others, passes into nasal cavity through
sphenopalatine foramen and crosses roof of nasal cavity to
septum. It passes downwards and forwards with its fellow
of the opposite side on the septum to the median incisive
foramina which open on to the incisive fossa, and ends in
the gums of the incisor teeth. Branches are given to the
mucous membrane of the septum.

(5) *Pharyngeal branch* that passes backwards through the
palatinovaginal canal to supply mucous membrane of upper part of nasopharynx and the opening into the sphenoidal sinus.

Mandibular nerve

This is the largest of the divisions of the trigeminal. The
large sensory root comes from the inferior angle of the
ganglion and is joined on its deep aspect near or in the
foramen ovale by the small motor root which lies deep to
the ganglion. The nerve leaves the skull by the foramen
ovale and divides into anterior and posterior divisions.

Branches from the trunk

Nervus spinosus: Accompanies middle meningeal artery
through foramen spinosum to supply the dura mater.

Nerve to medial pterygoid: Supplies medial pterygoid and
gives a branch to the otic ganglion.

Branches from anterior division
All motor except its continuation, the buccal which is sensory.

Masseteric: Passes laterally above lateral pterygoid, then through mandibular notch to deep surface of masseter; gives twigs to mandibular joint.

Deep temporal (2 or 3): Supply temporalis muscle; pass above lateral pterygoid muscle along greater wing of sphenoid and turn upwards deep to temporalis.

Buccal: Emerges between two heads of lateral pterygoid; passes on to buccinator and is sensory to mucous membrane and skin of cheek.

Lateral pterygoid: To lateral pterygoid muscle.

Branches from posterior division
All sensory except motor nerve to mylohyoid.

Auriculotemporal: Comes off by two roots between which the middle meningeal artery passes; runs backwards inferior to lateral pterygoid muscle then behind mandibular joint capsule; turns upwards with superficial temporal artery to temporal fossa where it becomes cutaneous; lies posterior to superficial temporal artery as they cross the zygomatic arch.

Gives off branches to temporomandibular joint, skin of upper half of outer surface of auricle, upper part of external acoustic meatus and tympanic membrane and scalp above meatus. Postganglionic parasympathetic fibres from the otic ganglion join the auriculotemporal nerve and go to the parotid gland.

Inferior alveolar (dental): Passes downwards medial to lateral pterygoid muscle, and posterior to the lingual nerve; then between the ramus of jaw and medial pterygoid muscle lateral to sphenomandibular ligament; enters mandibular foramen and runs in mandibular canal at first anterior to and then above the inferior alveolar artery; supplies molar and premolar teeth; divides into two branches, one (*incisor*) continues in the bone, and the other (*mental*) emerges in a backward direction from the mental foramen and supplies the skin of the lower lip from angle of mouth to midline, as well as the mucosa and gum of the anterior part of the vestibule.

Inferior alveolar gives off the *nerve to the mylohyoid* which descends in a groove on the medial side of ramus of jaw to lower surface of mylohyoid muscle, supplies it and gives off a branch to anterior belly of digastric.

Lingual nerve: Lies at first medial to the lateral pterygoid muscle; then passes downwards in front of inferior alveolar nerve, and is joined at an acute angle by the chorda tympani, a branch of the facial nerve. Thence it passes downwards and forwards between medial pterygoid and ramus of mandible inferior to superior constrictor of pharynx. Thence passes into mouth above posterior edge of mylohyoid grooving mandible just below crown of 3rd molar; runs anteriorly on surface of hyoglossus muscle on which muscle it lies lateral, then inferior, then medial to the submandibular duct.

The lingual nerve gives sensory branches to the submandibular ganglion and hypoglossal nerve. It contains fibres of general sensation from the mucous membrane of anterior two-thirds of tongue, floor of mouth and lingual gum. The chorda tympani which joins it contains fibres associated with taste from the anterior two-thirds of the tongue and preganglionic parasympathetic fibres which synapse in the submandibular ganglion.

Submandibular ganglion
Lies between hyoglossus muscle and deep part of submandibular gland. It receives:

(1) *Sensory fibres:* From the lingual whose cell bodies are in trigeminal ganglion.

(2) *Parasympathetic fibres:* Cell bodies are in superior salivatory nucleus; fibres run in nervus intermedius associated with facial nerve and leave in the chorda tympani (p. 260). Synapse in ganglion and postganglionic fibres supply sublingual and submandibular glands. Synapse for latter may be in gland itself.

(3) *Sympathetic fibres:* From plexus round facial artery. Cell bodies in superior and middle cervical ganglia.

Fibres from (1) and (3) pass through the ganglion and are distributed with the lingual nerve and to blood vessels.

Otic ganglion

Lies on the medial surface of the mandibular nerve on the tensor veli palatini close to the foramen ovale. It receives:

(1) *Parasympathetic fibres:* In lesser petrosal nerve whose cell bodies are in inferior salivatory nucleus; axons run in glossopharyngeal nerve which gives off *tympanic branch* to middle ear whence lesser petrosal nerve emerges and enters cranial cavity to go to foramen ovale. Synapse in ganglion and postganglionic secretomotor fibres go to parotid gland in auriculotemporal nerve.

(2) *Sympathetic:* From plexus on middle meningeal artery; cell bodies in superior cervical ganglion.

(3) *Motor:* Not present in other cranial ganglia. From nerve to medial pterygoid with cell bodies in motor nucleus of 5th nerve in pons, emerging from ganglion as nerves to tensor veli palatini and tensor tympani (both derived from the 1st pharyngeal arch).

ABDUCENT (6th CRANIAL) NERVE

Emerges from sulcus between pons and medulla near midline; cell bodies from nucleus in floor of 4th ventricle deep to facial colliculus.

Pierces dura mater on clivus and grooves tip of petrous bone to reach cavernous sinus in which it lies lateral to internal carotid; enters orbit by superior orbital fissure within fibrous ring between the heads of the lateral rectus above ophthalmic vein. Supplies lateral rectus.

FACIAL (7th CRANIAL) NERVE

Emerges laterally from between pons and medulla as two roots, large motor (medial) and smaller nervus intermedius (lateral). Motor nucleus is anterolateral to abducent nucleus.

Both roots pass forwards and laterally and enter the internal acoustic meatus in which they join. Nerve enters canal for facial nerve, first runs laterally above and between cochlea and vestibule of internal ear, then backwards in medial wall of middle ear just above fenestra vestibuli; as it bends backwards it has a swelling, the *facial (geniculate) ganglion;* finally it passes downwards in posterior border of medial wall medial to aditus to antrum (p. 311) to emerge from the bone at the stylomastoid foramen; it then passes laterally and forwards in the parotid gland dividing behind ramus of mandible into branches which further subdivide and intercommunicate.

Branches within the canal

Greater petrosal: Passes from ganglion of facial nerve through hiatus on anterior surface of petrous temporal then to anterior wall of foramen lacerum where it joins the deep petrosal to form the nerve of the pterygoid canal (p. 255). Contains only nervus intermedius fibres (preganglionic parasympathetic).

Nerve to stapedius (p. 311).

Chorda tympani: Given off just before exit from stylomastoid foramen, ascends and enters middle ear through posterior canaliculus for chorda tympani; it then passes between handle of malleus and long process of incus medial to tympanic membrane; leaves middle ear by anterior canaliculus for chorda tympani and emerges at the medial end of the petrotympanic fissure behind temporomandibular joint; grooves spine of sphenoid, then runs forwards between two pterygoids to join lingual nerve. It supplies mucous membrane of anterior two-thirds of tongue with taste fibres (cell bodies in facial ganglion) and preganglionic parasympathetic secretomotor fibres to submandibular sublingual and some buccal glands.

Branches at exit from stylomastoid foramen

Posterior auricular: Passes up behind external acoustic meatus and supplies auricular muscles and occipitalis of occipitofrontalis.

Nerve to stylohyoid muscle and posterior belly of digastric.

Branches on the face
These emerge from parotid gland and communicate with branches of trigeminal nerve.

Temporal: Supplies orbicularis oculi, frontalis of occipito-frontalis and corrugator.

Zygomatic: Supplies muscles of eyelid.

Buccal: Supplies buccinator, orbicularis oris and muscles of external nose.

Mandibular: Supplies muscles of lower lip and chin; commonly descends below angle of mandible and crosses inferior border of jaw where superficial to submandibular gland.

Cervical: Perforates cervical fascia beneath mandible, to supply platysma.

Nervus intermedius
Also incorrectly known as sensory root of facial nerve. Really a separate cranial nerve overlooked by ancients who counted only twelve pairs. Emerges between facial (motor) and vestibulocochlear nerves and contains (a) secretomotor (parasympathetic) fibres from superior salivatory nucleus, and (b) sensory (taste) fibres with cell bodies in facial ganglion and central processes which end in nucleus of tractus solitarius.

Principal branches are *greater petrosal* and *chorda tympani.*

VESTIBULOCOCHLEAR (8th CRANIAL) NERVE

Was called *auditory nerve.* Emerges between pons and inferior cerebellar peduncle in two parts—*cochlear* (hearing) and *vestibular* (equilibrium). *Cochlear ganglion* in modiolus of cochlea and *vestibular ganglion* in internal acoustic meatus. Central processes form the two nerves and go to *cochlear* and *vestibular nuclei* respectively in floor of 4th ventricle.

Nerves pass round inferior cerebellar peduncle with facial; all enter internal acoustic meatus.

Peripheral processes of cochlear ganglion cells end in relation to the *organ of Corti* in cochlea and those of vestibular ganglion in relation to the *hair cells* of the *cristae* and *maculae* of the *utricle, saccule* and *semicircular ducts.*

GLOSSOPHARYNGEAL (9th CRANIAL) NERVE

Emerges from the upper part of medulla in groove between olive and inferior cerebellar peduncle. There are no specific glossopharyngeal nuclei.

Passes laterally over flocculus of cerebellum to jugular foramen by which it leaves the skull with the inferior petrosal sinus in a separate tube of dura mater, in front of vagus and accessory nerves; makes a deep notch in petrous part of temporal bone (aqueduct of cochlea opens here). At exit and just below, there are two *ganglia,* the *superior* and *inferior.* The nerve then passes downwards and laterally between the internal carotid artery and jugular vein, and then forwards between the external and internal carotid arteries along the lateral border of the stylopharyngeus which it crosses superficially to pass on to the middle constrictor and deep to the hyoglossus to the tongue.

The *inferior ganglion* is connected by branches to (a) superior cervical ganglion of sympathetic, (b) auricular of vagus, and (c) superior ganglion of vagus.

The glossopharyngeal nerve gives off the following branches:

Meningeal: To inferior surface of tentorium cerebelli.

Tympanic: Arises from inferior ganglion and enters a minute canal in the bone between jugular foramen and carotid canal; reaches medial wall of middle ear and runs to the promontory where it forms a plexus with the facial; *lesser petrosal nerve* leaves the plexus and pierces the petrous temporal to reach the middle fossa just lateral to the hiatus for greater petrosal nerve; passes through the foramen ovale or foramen innominatum and ends in the otic ganglion (p. 259). The glossopharyngeal nerve is sensory to the lining of the middle ear and the lesser petrosal is a preganglionic parasympathetic nerve.

Pharyngeal: Sensory; joins pharyngeal of vagus and sympathetic to form *pharyngeal plexus* on middle constrictor; supplies mucous membrane of oropharynx.

Sinucarotid: Sensory to carotid sinus (baroreceptor) and carotid body (chemoreceptor).

Muscular: To stylopharyngeus.

Tonsillar: To palatine tonsil, oropharyngeal isthmus and part of soft palate.

Lingual: To the base of the tongue behind sulcus terminalis (posterior one-third) and also vallate papillae; general sensation and taste.

Sensory fibres have cell bodies in ganglia. Central processes of fibres of general sensation go to lateral part of dorsal nucleus of vagus and those of taste to the nucleus of tractus solitarius; cell bodies of preganglionic parasympathetic are in inferior salivatory nucleus; cell bodies of nerve to stylopharyngeus are in nucleus ambiguus.

VAGUS (10th CRANIAL) NERVE

Emerges from sulcus between inferior cerebellar peduncle and olive. Main nucleus is under vagal triangle on floor of 4th ventricle (p. 225); some fibres come from nucleus ambiguus of the medulla.

Passes from origin over the flocculus to jugular foramen through which it travels in same sheath as accessory and behind the glossopharyngeal nerve; in the foramen it has its *superior ganglion*. Just below foramen it receives the cranial root of the accessory and is enlarged by the *inferior ganglion*. Thence the nerve passes down in the carotid sheath, behind and between the artery and vein, to the root of the neck, where its course on each side of the body becomes different.

The *right nerve* passes between 1st part of subclavian artery and subclavian vein, then down side of trachea and behind right brachiocephalic vein to posterior part of the root of the right lung, forming the right *posterior pulmonary plexus*; from there, two cords run down on the oesophagus, communicate with nerve of opposite side (*oesophageal plexus*), join below into one trunk, which lies behind oesophagus and, passing through oesophageal orifice in diaphragm, is distributed to the posterior surface of the stomach (*posterior gastric nerve*); gives branches to coeliac and splenic plexuses.

The *left nerve* passes between the left subclavian and common carotid arteries and behind the left brachiocephalic vein, where the left phrenic crosses anterior to it; from there it passes to left of the arch of aorta to posterior surface of root of left lung, forming the left *posterior pulmonary plexus*; it then runs along anterior surface of oesophagus through diaphragm, to be distributed on the anterior surface of stomach (*anterior gastric nerve*); joins hepatic plexus.

Branches of vagus nerve

Meningeal: From superior ganglion, passes backwards to dura mater of posterior fossa (sensory).

Auricular: From superior ganglion, communicates with inferior ganglion of the glossopharyngeal, and enters a foramen between root of styloid process and jugular fossa; passes through temporal bone communicating with the facial nerve, and emerges through the auricular fissure just behind the external acoustic meatus to supply adjacent skin of meatus and auricle and part of tympanic membrane.

Pharyngeal: Principally formed by fibres from cranial root of accessory, passes between the two carotid arteries to the pharyngeal plexus. It is the motor nerve to the muscles of the pharynx (except stylopharyngeus) and muscles of soft palate (except tensor veli palatini).

Superior laryngeal: From inferior ganglion, passes downwards deep to internal carotid artery, where it divides into motor *external laryngeal nerve* (supplies the cricothyroid and cricopharyngeus muscles) and sensory *internal laryngeal nerve* which, passing through thyrohyoid membrane, is distributed to the mucous membrane of the piriform fossa and upper part of larynx (important nerve since it mediates the cough reflex).

Recurrent laryngeal: The *right nerve* arises in front of subclavian artery, winds below and behind it, and passes upwards behind common carotid and inferior thyroid arteries to right side of trachea. The *left nerve* arises to left of arch of aorta below which it winds to left of ligamentum arteriosum and passes upwards on left side of trachea. Each nerve ascends in the groove between trachea and oesophagus, and enters larynx by passing deep to lower border of inferior constrictor muscle. The nerve supplies all the muscles of the larynx (except cricothyroid), and gives a branch to cricopharyngeus.

Note: This nerve approaches the larynx by passing behind or between the terminal branches of the inferior thyroid artery.

Cervical cardiac (2 or 3): Superior are small; join cervical cardiac of sympathetic. Inferior arises just above 1st rib; the right runs along side of brachiocephalic artery and joins deep cardiac plexus; the left descends to left of arch of aorta and joins superficial cardiac plexus.

Thoracic cardiac: Right come from the trunk of the nerve; left from left recurrent laryngeal. Both end in deep cardiac plexus.

Bronchial (2 or 3): To anterior part of root of lung, joining with sympathetic to form the *anterior pulmonary plexus*. Numerous branches to posterior part of root of lung which join branches from 2nd, 3rd and 4th thoracic ganglia of sympathetic, forming *posterior pulmonary plexus*. Vagus is excitatory to smooth muscle of lung and secretomotor to mucous glands.

Oesophageal: Form oesophageal plexus in which both nerves intermingle; below level of lung roots.

Gastric: Right nerve is distributed to posterior part of stomach, and ends in the coeliac and splenic plexuses. Left supplies anterior surface and ends in the hepatic plexus. There is, however, some admixture of the two in the oesophageal plexus. Through the coeliac and superior and inferior mesenteric plexuses, the vagus supplies the alimentary tract as far as the left colic flexure by branches which travel with the arteries.

The vagus is *the* preganglionic motor parasympathetic nerve of the body. It also contains some taste fibres and some fibres to muscle derived from the pharyngeal arches. Most of the latter fibres come from the cranial accessory which joins the vagus.

Note: The branches of the vagus to the alimentary canal are excitor to the longitudinal muscle and at the same time relax the sphincters; this is in contrast with the cardiac branches, which are inhibitor.

ACCESSORY (11th CRANIAL) NERVE

Spinal root comes from upper five cervical segments and arises from the posterolateral part of the anterior horn of the cervical spinal cord; supplies the sternocleidomastoid and trapezius. The *cranial root* arises from the nucleus ambiguus and emerges from side of medulla; in line with the roots of the vagus and glossopharyngeal nerves; joins the vagus, and is distributed through that nerve to the pharyngeal, laryngeal and palatal muscles.

Note: The nucleus ambiguus is the motor nucleus for all muscles of pharynx, larynx and soft palate, except the tensor veli palatini.

The cranial root joins the spinal root inside the skull. The spinal root leaves the cord between the ventral and dorsal nerve roots and runs upwards behind ligamentum denticulatum to enter the skull through foramen magnum; it joins the cranial root to form the accessory nerve which passes to jugular foramen enclosed in the same sheath of dura mater as the vagus. In the jugular foramen the cranial root joins the vagus and the spinal root passes downwards between internal carotid artery and internal jugular vein, and then backwards superficial to the internal jugular vein to upper part of sternocleidomastoid which it pierces; at the same time it communicates with the branch to the muscle from the 2nd and 3rd cervical nerves. The spinal accessory crosses the occipital part of the posterior triangle and enters the deep surface of the trapezius where it joins with branches of the 3rd and 4th cervical nerves to form a plexus in the substance of the muscle. The spinal root is the motor nerve supply to the sternocleidomastoid and trapezius muscles; the branches from the cervical nerves are sensory.

HYPOGLOSSAL (12th CRANIAL) NERVE

Emerges by rootlets from groove between olive and pyramid. Nucleus is beneath hypoglossal triangle in floor of 4th ventricle (p. 225).

The nerve passes through hypoglossal canal in occipital bone, then downwards and forwards between vagus and accessory; then runs between internal carotid artery and internal jugular vein to the lower border of the digastric muscle; hooks round occipital artery, and crosses external and internal carotid arteries; is superficial to loop of lingual artery on middle constrictor; passes between the mylohyoid and hyoglossus muscles and ends by dividing into branches on the genioglossus.

Branches of hypoglossal nerve

Communicating: With the inferior ganglion of the vagus, superior cervical ganglion of the sympathetic, the loop between 1st and 2nd cervical nerves and lingual branch of the mandibular on the hyoglossus.

Meningeal: From C.1 (sensory); enters hypoglossal canal and supplies dura of posterior fossa.

Superior root of ansa cervicalis (descendens hypoglossi);
Slender branch given off as the nerve hooks round occipital artery, passes down over carotid sheath, and forms a loop (*ansa cervicalis*) with branch from 2nd and 3rd cervical (*descendens cervicalis or inferior root of ansa*); from this loop muscular branches are given to sternohyoid, sternothyroid and the two bellies of the omohyoid. This branch, and those to thyrohyoid and geniohyoid, do not arise from hypoglossal nucleus but are derived from 1st cervical spinal nerve.

Muscular: To intrinsic and extrinsic muscles of tongue (except palatoglossus); to thyrohyoid and geniohyoid (from 1st cervical nerve).

THE SPINAL NERVES

There are 31 pairs of spinal nerves: 8 cervical, 12 thoracic, 5 lumbar, 5 sacral and 1 coccygeal. Each nerve arises from the spinal cord by a *ventral (anterior)* motor and a *dorsal (posterior)* sensory *root*; the latter has a ganglion. The union of a dorsal and ventral root takes place in the intervertebral foramen from which it emerges and immediately divides again into *dorsal* and *ventral* (*anterior* and *posterior primary*) *rami*, each containing fibres from the two roots. The ventral rami supply the parts in front of the vertebral column, the dorsal rami the parts behind that column.

The *dorsal rami* of the spinal nerves are generally smaller than the anterior and pass directly backwards; each divides into a lateral and medial branch to supply the vertebral muscles and the skin for about 5 cm from the midline. The dorsal rami of the 1st cervical, the 4th and 5th sacral and the coccygeal nerves do not subdivide.

Each *ventral ramus* receives a grey communicating branch from the corresponding ganglion of the sympathetic trunk. The ventral rami from the 1st thoracic to the 2nd lumbar give white communicating branches to the sympathetic ganglia; the 2nd, 3rd and 4th sacral nerves give off branches which are parasympathetic.

THE CERVICAL NERVES

The dorsal rami

First cervical nerve
The dorsal ramus of the 1st cervical nerve passes backwards between the vertebral artery and the posterior arch of atlas and in the suboccipital triangle supplies the obliquus capitis inferior, the recti posteriores major and minor, the obliquus capitis superior, and the semispinalis capitis.

Second cervical nerve
Its medial branch is the *greater occipital nerve* which is specially large and supplies the skin of the posterior half of the scalp. It passes through the semispinalis capitis and trapezius and ascends with the occipital artery to the back of the scalp.

With the exception of the 1st cervical nerve, the *dorsal rami* of the cervical nerves divide into lateral branches which supply the muscles behind the vertebral column and medial branches which are larger than the lateral branches; they become cutaneous and supply the muscles through which they pass. The branch from the third cervical nerve supplies the skin at the base of the occiput (*third occipital*).

The medial branches from the 7th and 8th end in the muscles and have no cutaneous branches.

The ventral rami

The ventral rami of the first four cervical nerves form the *cervical plexus*, and those of the lower four, with part of that of the 1st thoracic nerve, compose the brachial plexus.

The ventral rami of the 1st and 2nd cervical nerves differ in their course from the rest. The ventral ramus of the 1st cervical nerve passes laterally on the posterior arch of the atlas to the lateral side of the lateral mass and passes forwards medial to the vertebral artery; it gives branches to the rectus capitis lateralis, longus capitis and rectus capitis anterior and joins the 2nd cervical nerve and hypoglossal nerve anterior to the atlas.

The ventral ramus of the 2nd cervical nerve winds forward around the atlanto-axial joint, and divides into an ascending branch which joins the 1st cervical, and a descending, which joins the 3rd cervical nerve.

THE CERVICAL PLEXUS

This is formed by the union of the ventral rami of the first four cervical nerves after each has received its grey ramus from the superior cervical ganglion. It is situated between the sternocleidomastoid and the scalenus medius muscles deep to the prevertebral fascia. Each nerve divides into an ascending branch which connects it with the nerve above, and a descending branch which joins it to the nerve below. From the loop between the 2nd and 3rd nerves cutaneous branches are given off to the head and neck, and from the union of the 3rd and 4th nerves superficial branches pass to the shoulder and chest, together with muscular and communicating branches.

Superficial branches

Transverse cervical (anterior cutaneous nerve of neck) (C.2, 3): From loop between 2nd and 3rd nerves, passes forward over the middle of the sternocleidomastoid, perforates the cervical fascia, and divides deep to platysma to supply skin of front of neck.

Great auricular (C.2, 3): Winds round posterior margin of sternocleidomastoid to reach parotid gland. Supplies skin of face over parotid gland, parotid fascia, skin of pinna and its lateral surface below external acoustic meatus, and skin over mastoid process.

Lesser occipital (C.2): Ascends along posterior border of sternocleidomastoid lying between ear and occipital artery to supply skin of scalp and also tip of cranial surface of auricle.

Supraclavicular (C.3, 4): Descends over posterior triangle and divides into medial (to neck above sternum), intermediate (skin above clavicle) and lateral (skin at tip of shoulder).

Deep branches
Communicating: From loop between 1st and 2nd to superior cervical ganglion and hypoglossal and vagus nerves.

Inferior root of ansa cervicalis (descending cervical nerve) (C.2, 3): Forms a loop with descending branch from hypoglossal nerve in front of carotid sheath (*ansa cervicalis*).

Muscular: To prevertebral muscles and through communication (C.1) with hypoglossal nerve to geniohyoid, thyrohyoid and other infrahyoid muscles.

Phrenic (C.4 mainly, also C.3 and 5): Lies in front of scalenus anterior; descends medially on it in front of subclavian artery and behind subclavian vein and enters the thorax, having crossed anterior to the internal thoracic artery at origin. In the thorax it descends in front of the root of the lung, lying between the pericardium and the mediastinal pleura, to the diaphragm, which it perforates; it is distributed on its abdominal surface. The right nerve lies to the right of right brachiocephalic vein and superior vena cava. Passes through caval orifice of diaphragm. The left nerve in the neck is crossed anteriorly by the thoracic duct and below crosses anterior to left vagus and runs downwards to left of arch of aorta. It is longer than the right. Sensory fibres from each supply the pericardium, pleura and diaphragmatic peritoneum.

Muscular: To levator scapulae (C.3, 4), scalenus medius (C.3, 4) and scalenus anterior (C. 4, 5).

The *branches to sternocleidomastoid* (C.2, 3) and *trapezius* (C.3, 4) are proprioceptive (sensory).

THE BRACHIAL PLEXUS

This is formed by the union and subsequent division of the ventral rami of the lower four cervical and part of the ventral ramus of the 1st thoracic nerves. The 5th and 6th cervical receive grey rami from the middle, and 7th and 8th from the inferior cervical ganglion. The following is the usual way in which the ventral rami unite and divide:

The 5th and 6th cervical join together at lateral border of the scalenus anterior to form an *upper trunk*. The 7th cervical forms a *middle trunk*. The 8th cervical and the 1st thoracic form a *lower trunk*.

Each of these trunks then subdivides into *anterior* and *posterior divisions*. The anterior divisions from the upper and middle trunks form the *lateral cord* of the plexus. The anterior division of the lower trunk forms the *medial cord* of the plexus. The posterior divisions of all the trunks unite to form the *posterior cord* (Fig. 59).

The ventral rami lie between scalenus anterior and medius and the trunks cross the posterior triangle. They are above the clavicle. Divisions lie behind the clavicle, and cords below the clavicle. Cords converge from above on axillary artery and surround its second part behind pectoralis minor. Here they lie medial, lateral and posterior to the artery, hence their names. Branches from the cords are related to third part of axillary artery.

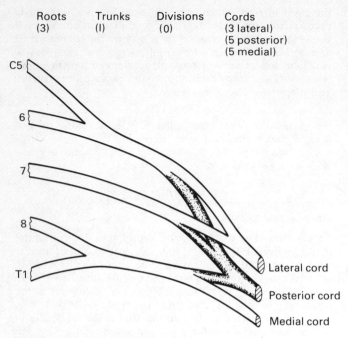

Fig. 59 Basic plan of brachial plexus.

Branches from the ventral rami

Dorsal scapular (nerve to the rhomboids): From C.5; passes through scalenus medius to medial border of scapula, then deep to levator scapulae, supplying it; ends deep to the rhomboid muscles. Runs with deep branch of transverse cervical artery.

Nerve to subclavius: Anteriorly from C.5, 6; passes downwards in front of the 3rd part of the subclavian artery and vein to the deep surface of the subclavius; often communicates with phrenic. This communication is the *accessory phrenic nerve* and passes into the thorax in front of the subclavian vein.

Long thoracic (nerve to serratus anterior): Posteriorly from C.5, 6, 7; 5th and 6th join in scalenus medius and leave its lateral border; joined in axilla by 7th, which passes in front of scalenus medius. Nerve descends behind midaxillary line deep to fascia on serratus anterior; supplies muscle segmentally (5th to upper two digitations, 6th to next two, 7th to last four).

Branch from the upper trunk

Suprascapular (C.5,6): Passes deep to trapezius to upper border of scapula and enters supraspinous fossa through suprascapular notch; gives off branches to the supraspinatus and an articular branch to the shoulder joint; thence it passes to the infraspinous fossa round lateral edge of scapular spine and ends in the infraspinatus.

Branches from the cords

The nerves given off are as follows:

Lateral cord	Medial cord	Posterior cord
Lateral pectoral	Medial pectoral	2 subscapulars
		Thoracodorsal
Musculocutaneous	Medial cutaneous nerve of upper arm	Axillary
Lateral head of median	Medial cutaneous nerve of forearm	Radial
	Ulnar	
	Medial head of median	

Lateral pectoral: From lateral cord, C.5, 6, 7, crosses over axillary artery and pierces clavipectoral fascia to reach deep surface of pectoralis major in which it communicates with medial pectoral.

Medial pectoral: From medial cord (C.8, T.1); passes between axillary artery and vein to pectoralis minor, which it pierces; ends in pectoralis major.

Subscapular: Two from posterior cord; *upper* from C.6; supplies upper part of subscapularis; *lower* from C.6, 7;

ends in teres major, having previously given a filament to subscapularis.

Thoracodorsal, C.6, 7, 8; runs along lower border of subscapularis with the subscapular vessels to supply the latissimus dorsi.

Axillary (circumflex): From posterior cord, C.5, 6; passes backwards with posterior circumflex vessels below subscapularis through quadrilateral space formed by teres major (below), teres minor (above), long head of triceps (medially), humerus (laterally); gives articular branch to shoulder joint and divides into an *anterior branch* which winds round neck of the humerus to supply deltoid and skin and a *posterior branch*, which gives a branch to teres minor and branches to deltoid and skin (*upper lateral cutaneous of arm*).

Note: Axillary nerve passes backwards inferior to shoulder joint and medial to surgical neck of humerus. It can be injured by downward dislocation at shoulder joint or fracture of neck of humerus.

Medial cutaneous nerve of upper arm: From medial cord (T.1); medial to axillary vein, communicates with intercostobrachial and descends along medial side of brachial vessels to middle of upper arm where it becomes cutaneous; supplies skin of medial side of upper arm as far as medial epicondyle. It communicates with the posterior branch of the medial cutaneous nerve of forearm.

Medial cutaneous nerve of forearm: From medial cord (C.8, T.1); lies in front of 3rd part of axillary artery, becomes cutaneous about middle of arm and divides into two branches. Anterior passes behind median basilic vein and supplies front of medial side of forearm as far as wrist. Posterior branch winds over medial epicondyle and supplies the back of medial side of forearm to about its middle.

Musculocutaneous: From lateral cord (C.5, 6, 7); perforates coracobrachialis, passes to lateral side of arm between biceps and brachialis; supplies all these muscles; may supply the elbow joint; becomes cutaneous as *lateral cutaneous nerve of forearm* just above elbow; passes behind median cephalic vein, divides into an anterior branch which passes along radial border of forearm and supplies ball of thumb, and a posterior branch which supplies skin of lower third of back of forearm on the radial side.

Median (C.5, 6, 7, 8, T.1): Arises by *lateral root* from lateral cord and *medial root* from medial cord; latter crosses 3rd part of axillary artery to join the lateral root. At first the nerve lies lateral to the axillary artery but about the middle of the upper arm it crosses anterior to the brachial artery to its medial side; it then passes between the two heads of pronator teres and deep to flexor digitorum superficialis; it continues straight down the forearm upon the flexor profundus, deep to the flexor superficialis to the deep surface of which it clings; at the wrist it lies deep or lateral to palmaris longus and between the tendons of the flexor superficialis and flexor carpi radialis. Passing deep to the flexor retinaculum, it becomes somewhat flattened and divides into two to supply the lateral 3½ digits.

Median nerve branches in forearm: Articular to elbow joint and muscular to pronator teres, flexor carpi radialis, palmaris longus, and flexor digitorum superficialis. It gives off the *anterior interosseous* just distal to the elbow joint; this nerve passes downwards lateral to the anterior interosseous artery on the membrane, between the flexor profundus and flexor pollicis longus, and ends in the deep surface of pronator quadratus. Supplies flexor pollicis longus, pronator quadratus, and lateral half of flexor digitorum profundus.

The *palmar cutaneous* branch is given off just above flexor retinaculum and ends in the skin of the palm.

Median nerve branches in hand: (a) *muscular*—supplies abductor brevis, opponens, and flexor brevis of thumb. This branch comes off deep to flexor retinaculum and curves backwards and laterally around lower border of flexor retinaculum; and (b) *palmar digital*—five in number, supplying lateral 3½ digits. Nerves arise deep to arteries in palm but become anterior to them before entering the digits. 1st and 2nd supply the thumb; 3rd to radial side of index finger also supplies 1st lumbrical; 4th to adjacent sides of index and middle fingers and supplies 2nd lumbrical; 5th supplies adjacent sides of ring and middle fingers and joins a branch of the ulnar; sometimes gives a branch to the 3rd lumbrical.

Ulnar (C.8, T.1): From medial cord (also by a lateral head from C.7 in 95% of individuals for supply of flexor carpi ulnaris); passes down the medial side of axillary and brachial arteries to middle of upper arm; passes backwards with ulnar collateral artery through medial intermuscular septum and downwards to groove between olecranon and medial epicondyle. Thence it passes forwards into forearm between the two heads of flexor carpi ulnaris and descends under cover of that muscle, along ulnar side of forearm medial to ulnar artery, as far as the pisiform bone; it then runs anterior to the flexor retinaculum lateral to that bone and medial to hook of hamate into palm, where it divides into *superficial* and *deep palmar branches.*

Ulnar nerve branches in forearm: Articular to elbow and wrist joints and *muscular* to flexor carpi ulnaris and medial half of flexor digitorum profundus. The *palmar cutaneous* arises near middle of forearm, and accompanies ulnar artery into hand; supplies skin of medial side of the palm; joins the palmar cutaneous of median. The *dorsal cutaneous* comes off about 8 cm above pisiform bone, winds round ulna deep to flexor carpi ulnaris and supplies medial side of little finger, and adjacent sides of ring and little fingers on their dorsal aspects.

Ulnar nerve branches in hand: Superficial branch supplies palmaris brevis and ends in two digital branches for medial 1½ digits, the lateral communicating with the median. *Deep terminal branch* accompanies deep palmar arch; supplies the small muscles of the hypothenar eminence (flexor brevis, abductor and opponens of little finger), all the palmar and dorsal interossei and the medial two lumbricals. In the space between the thumb and index finger the nerve ends by supplying the adductor pollicis and 1st palmar interosseus.

Radial (C.5, 6, 7, 8, T.1): From posterior cord; lies posterior to axillary and brachial arteries and passes between long and medial heads of triceps brachii; then winds between lateral and medial heads of triceps in the spiral groove of humerus with profunda brachii artery to the lateral side of arm. Pierces the lateral intermuscular septum and passes to the front of the lateral epicondyle where it lies between the brachioradialis and brachialis; divides into *superficial* and *deep branches* (former was called *radial nerve* and latter the *posterior interosseous nerve*).

Radial nerve branches in upper arm: Muscular to the three heads of triceps, anconeus, brachioradialis, extensor carpi radialis longus and usually brachialis. *Posterior cutaneous nerve of arm* comes off in axilla, supplies skin on back of arm to near olecranon; *lower lateral cutaneous nerve of arm* perforates lateral head of triceps, accompanies cephalic vein to elbow and supplies skin of lower half of upper arm in front; *posterior cutaneous nerve of forearm* usually comes off with preceding nerve and supplies skin of lower part of upper arm, and back part of radial side of forearm as far as the wrist.

Radial nerve branches in forearm: Deep branch (posterior interosseous) passes backwards between two layers of supinator to go to back of forearm where it passes between the superficial and deep layers of muscles to about middle of forearm; it then passes deep to extensor pollicis longus to reach the interosseous membrane on which it lies as far as the wrist; there it ends and gives branches to the ligaments etc. Supplies supinator, extensors carpi radialis brevis, digitorum, digiti minimi, carpi ulnaris, pollicis longus and brevis, and indicis and abductor pollicis longus.

The *superficial branch (radial nerve)* passes downwards lateral to radial artery, under cover of brachioradialis, to within 8 cm of lower end of radius, where the nerve passes backwards deep to tendon of brachioradialis; becomes cutaneous by piercing the fascia on lateral side of forearm and divides into two branches; *lateral* supplies ball and lateral border of thumb, joining with the lateral cutaneous nerve of forearm; *medial* joins lateral and posterior cutaneous nerves of forearm, and dorsal cutaneous branch of ulnar. It gives off four *dorsal digital nerves*, 1st to medial side of thumb, 2nd to lateral side of index, 3rd to adjacent sides of index and middle, 4th to adjacent sides of middle and ring fingers (does not supply skin beyond distal interphalangeal joint).

THE THORACIC NERVES

These are twelve in number. The 1st comes out between the 1st and 2nd thoracic vertebrae and its greater part joins the brachial plexus. The 12th thoracic spinal nerve emerges from between the 12th thoracic and 1st lumbar vertebrae. Each nerve at its exit from the intervertebral foramen divides into ventral and dorsal rami. The 1st and 12th nerves, however, require a separate description.

The *dorsal rami* pass backwards between the transverse processes and divide into lateral and medial branches which supply the vertebral muscles of the back. Cutaneous branches are derived as follows: the six upper come from the medial branches and the six lower from the lateral.

The *ventral rami* form the *intercostal nerves*. There are eleven on each side (that of the 12th is called the *subcostal nerve*); each gives a white ramus to the corresponding sympathetic ganglion and receives a grey ramus from it.

The upper six intercostal nerves pass forwards (Fig. 60) in the intercostal spaces inferior to the vessels, lying at first between the pleura and posterior intercostal membrane, then between the innermost and internal intercostal muscles; as they approach the front of the thorax they lie between the internal intercostal muscle and the pleura, extend forwards to the sternum, and cross anterior to the internal thoracic artery; they end as the *anterior cutaneous nerves* of the thorax by perforating the intercostal muscles and pectoralis major.

Each intercostal nerve gives off:

Collateral branch which arises posteriorly and runs in neurovascular plane in lower part of space; supplies intercostal muscles, parietal pleura and periosteum of ribs;

lateral cutaneous which comes off midway between head of rib and sternum. (The 1st intercostal nerve has no lateral cutaneous branch.) Each branch pierces the intercostal muscles, and divides into anterior and posterior branches, which supply the mammary gland and skin. The lateral cutaneous branch of the 2nd nerve, or *intercostobrachial*, does not divide; it crosses the axilla, joins the medial

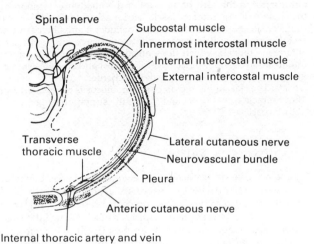

Fig. 60 Diagram of an intercostal space showing course of an intercostal nerve.

cutaneous nerve of arm and supplies the skin of the medial side of the arm.

The lower six intercostal nerves pass like the upper ones to the front of the intercostal spaces and then between the internal oblique and transversus abdominis to the sheath of the rectus; they pierce the posterior layer of the sheath and pass medially behind the rectus abdominis which they perforate to terminate near the midline as anterior cutaneous branches.

The *lower intercostal nerves give off:*

Collateral branches as above and *lateral cutaneous branches* which supply the skin of the abdomen and have anterior and posterior branches.

Note: Most of the ventral ramus of the 1st thoracic spinal nerve goes to the brachial plexus which it reaches by passing obliquely across the neck of the 1st rib, but a small branch is given off to the 1st intercostal space (it has no lateral cutaneous branch). The ventral ramus of the 12th nerve does not lie in an intercostal space, but inferior to the 12th rib in front of the quadratus lumborum; it then pierces the lumbar fascia and passes forwards between transversus abdominis and internal oblique; it ends by perforating rectus abdominis; it is remarkable for the large size of its lateral cutaneous branch, which does not divide, but, piercing internal and external oblique, passes over iliac crest and supplies skin over gluteal region as far as the greater trochanter.

Note: The skin in the region of the umbilicus is innervated by the 10th thoracic spinal nerve and that of the suprapubic region by the 12th thoracic nerve.

THE LUMBAR NERVES

These are five on each side. The *dorsal rami* pass backwards between the transverse processes and divide into medial and lateral branches. The medial branches end in the muscles; the lateral also supply muscles and the upper three give large cutaneous branches to gluteal region.

The *ventral rami* increase in size from above downwards; the upper two give off white rami and all five receive grey rami from the corresponding sympathetic ganglia; the upper three ventral rami and greater part of the 4th form the *lumbar plexus*; the rest of the 4th joins with the 5th to form the *lumbosacral trunk* which joins the sacral plexus.

THE LUMBAR PLEXUS

This is formed in the psoas by the ventral rami of the upper four lumbar nerves in the following manner:

The 1st gives off the iliohypogastric, the ilio-inguinal, a branch to the genitofemoral and a communicating branch to the 12th thoracic and 2nd lumbar.

The 2nd gives off branches to the genitofemoral and lateral cutaneous of thigh, and a communicating branch to the 3rd.

The 3rd contributes to the lateral femoral cutaneous, the femoral, the obturator.

The 4th contributes to the femoral and the obturator, and gives off a communicating branch to the 5th.

Muscular branches are supplied to the psoas and quadratus lumborum.

Branches of lumbar plexus

Iliohypogastric (L.1): Appears at upper part of lateral border of psoas, passes anterior to quadratus lumborum to iliac crest, and, piercing the transversus abdominis, divides into (a) lateral branch which pierces the two oblique muscles, crosses iliac crest behind lateral cutaneous of 12th thoracic nerve and supplies skin of buttock, and (b) anterior branch which pierces internal oblique near anterior superior iliac spine and then external oblique aponeurosis above superficial inguinal ring to end in skin above pubis.

Ilio-inguinal (L.1): Passes anterior to quadratus lumborum and iliacus to iliac crest and pierces transversus abdominis and internal oblique; it then runs parallel to and just above inguinal ligament in the inguinal canal where it accompanies the cord through superficial inguinal ring, and is distributed to the skin of the groin and part of the scrotum.

Genitofemoral (L.1, 2): Passes on the psoas towards inguinal ligament and divides into *genital branch*, which crosses external iliac artery, enters inguinal canal through deep inguinal ring, accompanies spermatic cord, and supplies the cremaster muscle (in the female it accompanies the round ligament of the uterus) and *femoral branch* which passes deep to inguinal ligament, perforates fascia on lateral side of femoral artery near saphenous opening and supplies small area of skin of upper and front part of thigh.

Lateral femoral cutaneous (L2, 3): Perforates lateral border of psoas, runs obliquely across iliacus and enters thigh through fibrous sling beneath lateral part of inguinal ligament about 2.5 cm medial to anterior superior spine; supplies anterolateral surface of upper half of thigh.

Obturator (L.2, 3, 4, ventral divisions): Emerges from medial border of psoas, near brim of pelvis, above obturator artery and below external iliac, runs round lateral wall of pelvis 1 cm below brim to canal in upper part of obturator foramen. In this canal it divides into (a) *anterior branch* which descends in front of adductor brevis, behind pectineus and adductor longus; it supplies the hip joint, gracilis, adductor longus, adductor brevis, femoral artery, and a cutaneous branch to plexus under sartorius, sometimes supplies pectineus; communicates with accessory obturator when this is present. Supplies skin above adductor tubercle (lower medial part of thigh); (b) *posterior branch* which passes through obturator externus and behind adductor brevis; supplies a large branch to adductor magnus, and gives branches to obturator externus, adductor brevis when the latter is not supplied by the anterior branch, and a branch which accompanies the popliteal artery and goes to the knee joint.

Accessory obturator (L3, 4): When present it passes down on medial side of psoas, over superior ramus of pubis, under pectineus, and supplies pectineus and hip joint; communicates with anterior branch of obturator.

Femoral (L.2, 3, 4 dorsal divisions): Emerges from lower part of lateral border of psoas, and descends between that muscle and the iliacus, lying on the lateral side of the external iliac vessels. It supplies the iliacus and femoral artery whilst in the pelvis. Enters thigh deep to the inguinal ligament and lateral to femoral sheath; divides into branches, between which the lateral circumflex artery passes laterally.

Muscular branches:
   (1) *To pectineus:* Generally two, which pass medially under femoral vessels to muscle.

(2) *To sartorius:* Given off with intermediate femoral cutaneous.

(3) *To rectus femoris:* Also supplies hip joint

(4) *To vastus lateralis:* Accompanies descending branch of the lateral circumflex artery, and gives an articular branch to the knee joint.

(5) *To vastus medialis:* Passes into adductor canal and accompanies the deep branch of the descending genicular artery; gives off an articular branch to the knee joint.

(6) *To vastus intermedius:* Two or three, the medial one supplying articularis genus and knee joint.

Cutaneous branches:

(1) *Intermediate femoral cutaneous:* Pierces fascia lata about 8 cm distal to the inguinal ligament and divides into two branches to supply the skin of the front of the thigh as far as the knee; gives a branch to sartorius.

(2) *Medial femoral cutaneous:* Passes obliquely across to medial side of femoral artery, and pierces fascia lata in lower third of thigh; supplies skin of the lower third of medial side of thigh and skin of the leg; communicates in the thigh with the obturator and the saphenous nerves, forming in adductor canal the *subsartorial plexus.*

(3) *Saphenous:* Accompanies femoral artery, lying on its lateral side as far as adductor canal in which it crosses artery; leaves canal at its lower end by passing medially deep to sartorius. Here it becomes subcutaneous and continues with the great saphenous vein along medial side of leg, behind medial border of tibia, and in front of medial malleolus; distributed on medial side of foot as far as ball of great toe. In its course it gives off a branch to plexus deep to sartorius formed by obturator and medial femoral cutaneous nerves, to patellar plexus, and below the knee to the skin on the anterior and medial surfaces of the leg (*infrapatellar branch*).

(4) *Patellar plexus:* This is formed by communications between branches of the medial, intermediate and lateral femoral cutaneous nerves, together with the infrapatellar branch of the saphenous nerve.

THE SACRAL NERVES

These are five on each side. The roots of origin are in the cauda equina (p. 216), and in this region the *spinal* or *dorsal root ganglia* are inside the vertebral canal, though outside the dura mater. Each nerve divides into ventral and dorsal rami.

The *dorsal rami* of the upper four emerge from the posterior sacral foramina and the fifth at the lower end of the vertebral canal; the upper three divide into medial and lateral branches, the former supplying the multifidus, the latter the skin over sacrum, coccyx and posterior gluteal region; the two lower nerves do not divide, and supply filaments to skin over coccyx; the 5th communicates with the coccygeal nerve.

The *ventral rami* decrease in size from above downwards. The upper four issue from the anterior sacral foramina, the 5th emerging between sacrum and coccyx. Each nerve receives a grey ramus from the sympathetic. The 1st to the 4th nerves together with the lumbosacral trunk (L.4, 5) form the sacral plexus, whilst branches of the 4th and 5th enter the coccygeal plexus. Pelvic parasympathetic branches issue from the 2nd, 3rd and 4th sacral nerves.

THE SACRAL PLEXUS

Lies on anterior surface of piriformis, behind the internal iliac artery and the rectum and external to the pelvic fascia.

Branches of sacral plexus

   Muscular:

      To piriformis (S.1, 2).

      Nerve to obturator internus (L.5, S.1, 2): To obturator internus and gemellus superior; emerges from pelvis through greater sciatic foramen, winds over ischial spine lateral to the internal pudendal artery; passes into perineum through lesser sciatic foramen to medial surface of muscle.

      Nerve to quadratus femoris (L.4, 5, S.1): To quadratus

femoris and gemellus inferior; passes on bone deep to gemelli and obturator internus tendon to deep surface of quadratus; gives an articular branch to hip joint.

Superior gluteal (L.4, 5, S.1): Passes out of greater sciatic foramen above piriformis with the superior gluteal vessels; divides into a superior branch, which passes between the gluteus medius and minimus and supplies the medius, and an inferior branch supplying gluteus minimus and tensor fasciae latae.

Inferior gluteal (L.5, S.1, 2): Passes out of pelvis inferior to piriformis at the lower border of which it runs backwards; divides into numerous branches and enters deep surface of gluteus maximus towards its medial attachment.

Cutaneous:

Posterior femoral cutaneous (S.2, 3, 4): A cutaneous nerve to lower part of buttock, perineum and back of thigh; passes below piriformis with the inferior gluteal artery and runs downwards deep to gluteus maximus and down the back of the thigh deep to fascia lata which it pierces in the popliteal space.

It gives off (a) Perineal branch: to skin of upper and medial side of thigh and scrotum (labia); joins the lateral scrotal (labial) branch of pudendal; (b) Gluteal branches: wind round gluteus maximus and supply skin over lower part of buttock. (c) Femoral branch: to skin of thigh, popliteal region, and upper part of calf of leg.

Perforating cutaneous (S.2, 3): Perforates sacrotuberous ligament and lower border of gluteus maximus and supplies skin over lower part of buttock.

Sciatic: It is derived from the lumbosacral trunk and the 1st, 2nd and 3rd sacral nerves and is the largest nerve in the body; it is the main continuation of the sacral plexus and lies on piriformis. It passes out of pelvis below the piriformis, and between the tuberosity of ischium and greater trochanter, resting in turn upon the ischium, obturator internus, quadratus femoris and adductor magnus. It enters thigh deep to hamstrings, and is accompanied by the inferior gluteal artery which supplies a branch to its substance (*companion artery of sciatic nerve*). At a variable distance between the sacral plexus and lower part of the thigh, but generally about its middle, the nerve bifurcates into *tibial (medial popliteal)* and *common peroneal (lateral popliteal)*. In cases of division within the pelvis the common peroneal nerve pierces piriformis. Tibial nerve is derived from anterior divisions (L.4, 5, S.1, 2, 3) and common peroneal from posterior divisions of L.4, 5, S.1 and 2. The sciatic nerve supplies the back of the thigh and structures distal to the knee.

The sciatic nerve in the thigh gives branches to semimembranous, semitendinosus, both heads of biceps femoris, and ischial part of adductor magnus. Branch to short head of biceps is derived from common peroneal, the others from tibial component of sciatic nerve.

Tibial (medial popliteal): Larger terminal branch; passes down middle of popliteal space to lower border of popliteus, where it formerly took the name of *posterior tibial*; it is at first superficial and lateral to the artery, but at the lower end of the space, deep to gastrocnemius, crosses to medial side.

It gives off the following branches: (a) *Articular* to the knee joint (3): These accompany the medial superior, medial inferior and middle genicular arteries respectively. (b) *Muscular* in the popliteal fossa: To the gastrocnemius, one for each head, the lateral one also supplying the plantaris; to the soleus and to the popliteus, the latter turning round lower border of muscle and entering its anterior (deep) surface. (c) *Sural:* Arises in popliteal fossa and passes down leg superficially between two heads of the gastrocnemius; pierces the deep fascia about middle of leg where it is joined by the sural communicating branch of the common peroneal; it accompanies the small saphenous vein round the lateral malleolus and supplies skin of lateral side of foot and little toe, communicating with superficial peroneal on dorsum of foot.

Beyond the lower border of the popliteus, the tibial nerve accompanies the posterior tibial vessels (p. 182) on the deep muscles deep to soleus to midway between medial malleolus and heel, where it divides into *lateral* and *medial plantar nerves*. It is at first medial to the artery, but afterwards crosses to the lateral side.

In the leg the tibial nerve gives off *branches* to soleus, tibialis posterior, flexor digitorum longus, and flexor hallucis longus, the last accompanying the peroneal artery. It also gives off the *medial calcanean* which pierces flexor retinaculum to supply skin of heel and medial side of sole of foot.

Medial plantar: Larger terminal branch of the tibial; accompanies medial plantar artery along medial side of foot (the larger nerve thus accompanies the smaller artery); corresponds in distribution with median nerve of hand; passes between the abductor hallucis and flexor digitorum brevis to divide opposite the bases of the metatarsal bones into four branches, the most lateral of which communicates with the lateral plantar.

It gives off cutaneous branch to medial side of sole and muscular to abductor hallucis and flexor digitorum brevis. It also supplies the tarsal and metatarsal joints. In the sole it gives off four *plantar digital nerves*. The 1st supplies medial border of 1st toe and the flexor hallucis brevis, the 2nd supplies the adjacent sides of the 1st and 2nd toes and the 1st lumbrical, the 3rd supplies the adjacent sides of the 2nd and 3rd toes, and the 4th supplies the adjacent sides of the 3rd and 4th toes and joins a branch from the lateral plantar.

Lateral plantar: Corresponds in distribution to ulnar nerve in hand; passes across to lateral side of foot with lateral plantar artery between 1st and 2nd layers of muscles (p. 184) and supplies the abductor digiti minimi and flexor digitorum accessorius. At the lateral border of the latter muscle it divides into two branches: (a) *superficial* which gives off two *plantar digital nerves,* one supplying the lateral side of the little toe, the flexor digiti minimi brevis, and the two interossei of the 4th space; the other supplies the adjacent sides of the 4th and 5th toes and communicates with the medial plantar; and (b) *deep* or muscular which accompanies deep part of plantar artery passing medially between 3rd and 4th layers of muscles; supplies adductor hallucis (oblique and transverse heads), three lateral lumbricals and interossei of three medial spaces.

Common peroneal (lateral popliteal) nerve: Passes downwards in the popliteal space under cover of and medial to biceps tendon, then over lateral head of gastrocnemius to the fibula; winds forwards round the neck of that bone where it is superficial and palpable and frequently injured, and pierces the peroneus longus; in that muscle divides into *deep peroneal (anterior tibial)* and *superficial peroneal (musculocutaneous) nerves.*

It gives off two *articular* branches to the knee (generally given off together and accompany lateral superior and inferior genicular arteries); *lateral cutaneous nerve of the calf* which supplies skin of back and lateral side of leg in upper third; and *sural communicating* which arises close to head of fibula and joins the sural of tibial nerve.

Deep peroneal (anterior tibial): Passes to front of interosseous membrane by piercing extensor digitorum longus to reach lateral side of anterior tibial artery, with which it descends on interosseous membrane to the ankle joint; it then divides into a medial and lateral branch; in middle one-third of leg it lies in front of artery, and in lowest one-third it is again lateral to it.

It gives off (a) *recurrent articular* to knee, which accompanies anterior tibial recurrent artery to joint; (b) *muscular branches* to tibialis anterior, extensor digitorum longus, peroneus tertius and extensor hallucis longus; (c) *lateral branch* which passes laterally deep to extensor digitorum brevis; supplies this muscle and the joints of the tarsus and metatarsus; and (d) *medial branch* which accompanies dorsalis pedis artery to 1st metatarsal space, lying lateral to it; supplies adjacent sides of 1st and 2nd toes, communicating with the superficial peroneal.

Superficial peroneal (musculocutaneous): Supplies peroneus longus and brevis, and cutaneous branches to dorsum of foot. It passes down between peronei and the long extensor of toes, piercing deep fascia at lower one-third of leg.

Branches: muscular to peroneus longus and peroneus brevis; *cutaneous* to lower part of leg; *medial* passing over ankle to medial side of 1st toe and adjacent sides of 2nd and 3rd toes: communicates with saphenous and deep peroneal (anterior tibial) nerves; and *lateral* supplying adjacent sides of 3rd, 4th, and 5th toes; communicates with sural.

Pudendal (S.2, 3, 4): Passes out of greater sciatic foramen between piriformis and coccygeus, medial to sciatic nerve; winds over sacrospinous ligament medial to internal pudendal artery, and enters perineum through the lesser sciatic foramen on medial side of internal pudendal artery; it then enters, with accompanying vessels, the pudendal canal in the obturator fascia on the lateral wall of the ischiorectal fossa and divides into its three branches.

(1) *Inferior rectal:* Passes medially in ischiorectal fossa to supply external anal sphincter and skin of anus; communicates with perineal branch of posterior femoral cutaneous and scrotal nerves.

(2) *Perineal:* Largest terminal branch, accompanies artery, and divides into:

Scrotal (labial) two in number: medial passes with scrotal artery either under or over the superficial transversus perinei to supply the scrotum; the lateral gives a branch to the anus, and piercing the membranous layer of the superficial fascia, supplies the scrotum (labia).

Muscular branches supply transverse perineal muscles, ischiocavernosus, bulbospongiosus, sphincter urethrae, external anal sphincter and levator ani.

Nerve to bulb: Pierces bulbospongiosus and supplies bulb of corpus spongiosum.

(3) *Dorsal nerve of the penis (clitoris):* Accompanies internal pudendal artery deep to perineal membrane lying lateral to the artery; pierces the perineal membrane and the suspensory ligament to reach dorsum of penis (clitoris) along which it runs as far as the glans, gives off many branches to supply the organ, and joins branches of the sympathetic.

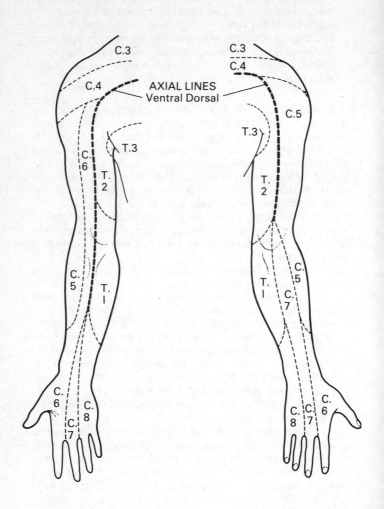

Fig. 61a Dermatomes of upper limb.

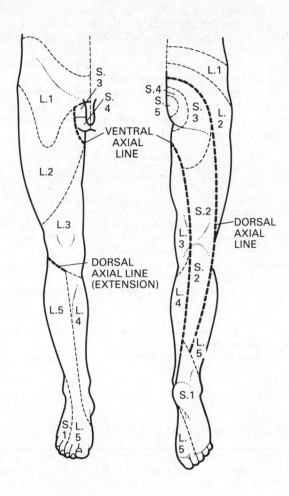

Fig. 61b Dermatomes of lower limb.

COCCYGEAL PLEXUS

Perineal branch of fourth sacral: Leaves pelvis between coccygeus and levator ani, supplying each on its pelvic surface. In perineum supplies external anal sphincter, peri-anal skin and dartos muscle.

The *ventral ramus of the 5th sacral* emerges from sacral hiatus and turning forwards pierces coccygeus. It is joined by a small branch from the 4th sacral and *ventral ramus of coccygeal nerve* to form *coccygeal plexus*. From this the *coccygeal nerve* arises which pierces the sacrotuberous ligament to supply the skin over the coccyx.

DERMATOMES

The skin of the trunk is supplied segmentally by the intercostal nerves. In the limbs there is also a segmental supply. The area of skin supplied by one spinal nerve is called a *dermatome*. The arrangement of the limb dermatomes is shown in Fig. 61. Note that at the *anterior* and *posterior axial lines* discontinuous dermatomes lie in contact with each other and that the dermatome of the central spinal cord segment of each plexus lies at the extremity of the limb (C.7 for hand, S.1 for foot). Such peripheral dermatomes on the limb have no connection with skin of body wall. Adjacent dermatomes overlap considerably but there is no overlap across the axial lines.

Note: Exact delineation of dermatomes remains uncertain and any chart is only approximate.

THE AUTONOMIC NERVOUS SYSTEM

This is the visceral (splanchnic) part of the nervous system and comprises the nerve supply to the viscera, glands and blood vessels, and to non-skeletal muscle in general throughout the body. It consists of a *sympathetic* part and a *parasympathetic* part. All the efferent fibres of the sympathetic arise in the thoracic and upper lumbar segments of the spinal cord (T.1–L.2); the parasympathetic fibres are incorporated in some of the cranial and sacral (S.2, 3, 4) nerves. All the efferent fibres of both systems have a common feature—they are interrupted in their course by a synapse in a peripheral ganglion. The *ganglia of the sympathetic* form a series along the sides of the bodies of the vertebral column; the *ganglia of the parasympathetic* are less well defined and consist of ganglia or groups of cells in or

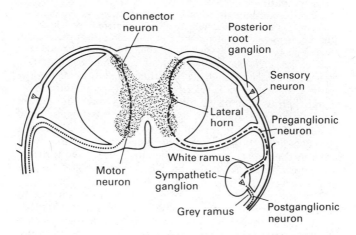

Fig. 62 Arrangement of neurons of somatic reflex arc (left) and neurons of sympathetic system (right).

near the walls of the viscera. In addition to these anatomical differences, the two systems differ functionally, being usually antagonistic to one another in their physiological and pharmacological actions; the vagus (parasympathetic), for example, slows the heart and the sympathetic accelerates it.

The fibres which leave the central nervous system are designated as *preganglionic*, and those which leave the ganglia as *postganglionic*. The system therefore consists of preganglionic fibres, ganglia and postganglionic fibres.

These fibres may be regarded as analogous to the component fibres of a simple reflex arc (Fig. 62). The connector cell (internuncial neuron) of the spinal reflex becomes the connector cell of the autonomic system with its cell body in the lateral horn of grey matter of the spinal cord for the sympathetic system and in the Edinger-Westphal (III), salivatory (VII and IX) and vagal (X) nuclei of the brain stem for the cranial part and the lateral horn of the 2nd to 4th sacral spinal cord segments for the sacral part of the parasympathetic system.

Note: It is not enough to know the distribution of the autonomic system; it is also essential to know the situation of the cell bodies of the fibres under consideration. Think in complete neurons, not merely in axons. This is true also of the central nervous system.

The *preganglionic fibres*, which are myelinated, pursue a course of variable length within the spinal or cranial nerve with which they are associated, and, leaving the nerve, make their way to the ganglion or ganglia to which they are destined. In the case of the spinal nerves, the preganglionic fibres leave the ventral rami (mostly the intercostal nerves) and constitute the *white rami communicantes*. They arise from cells in the lateral column of the cord, and leave through the ventral roots from the 1st thoracic to the 2nd lumbar nerves (Fig. 62). They continue into the ventral rami.

The *postganglionic fibres* conduct the nervous impulse to its final destination, and usually they are non-myelinated. In the case of the spinal nerves, they constitute the *grey rami communicantes*. The latter usually pass back into the spinal nerve and are thus distributed with it. Other postganglionic fibres go to the viscera either directly or by way of the blood vessels.

The ganglia of the autonomic system are in three groups: (1) ganglia of the sympathetic nerve trunk, (2) collateral ganglia or plexuses, (3) ganglia in or near the walls of the structures supplied.

The sympathetic preganglionic fibres synapse in either the ganglia of the sympathetic trunk or the collateral ganglia, and the parasympathetic preganglionic fibres usually synapse in (3).

The ganglionated sympathetic trunk is placed one on each side of the vertebral column, extending from the base of the skull to the pelvis. Each ganglionated chain comprises three cervical, eleven thoracic, four lumbar and four sacral ganglia.

The ganglia from the 1st thoracic to the 2nd lumbar receive from the corresponding spinal nerves white communicating branches. The white communicating branches given off from the 2nd, 3rd and 4th sacral nerves belong to the parasympathetic. To every spinal nerve, without exception, there is given off a grey communicating branch. The postganglionic fibres which constitute this branch are thus distributed with the spinal nerve. Numerous preganglionic fibres with their synapses in the collateral ganglia pass without interruption through the ganglionated trunk and form the *splanchnic nerves*.

From the cervical ganglia which receive their preganglionic branches from the 1st and 2nd thoracic spinal cord segments, grey branches are given off to the cervical nerves. Postganglionic fibres also go to the dilator pupillae and to the smooth muscle in the upper eyelid and at the back of the orbit. They also go to the buccal, salivary and lacrimal glands by way of the plexuses along the carotid vessels and their branches. These fibres are mainly vasoconstrictor.

The collateral ganglia form various plexuses—in the thorax, the cardiac, pulmonary and oesophageal, and in the abdomen, the coeliac, aortic and hypogastric plexuses.

The parasympathetic ganglia lie in or near the viscera supplied. A summary of the distribution of the sympathetic and parasympathetic is given in the appended scheme (pp. 305–306).

The cell stations of the parasympathetic fibres contained in the oculomotor, facial and glossopharyngeal nerves are found in the cranial autonomic ganglia.

Cranial autonomic ganglia

In the parasympathetic system preganglionic neurons usually synapse with postganglionic in the walls of the viscus. In the head they are grouped into four ganglia: ciliary (p. 253), pterygopalatine (p. 255), submandibular (p. 258) and otic (p. 259). Only parasympathetic fibres synapse in these ganglia. Each ganglion receives a sensory and sympathetic branch except the otic which receives a motor

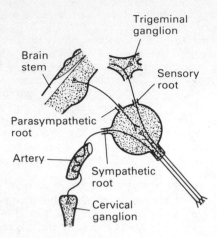

Fig. 63 Basic plan of cranial autonomic ganglia.

branch from the nerve to the medial pterygoid muscle for tensor tympani and tensor veli palatini muscles. All these fibres pass through the ganglia without synapsing. The study of the four ganglia is simplified if no nerve fibre is ever thought of without visualizing the site of its cell body (Fig. 63). In all cases the sensory fibres have their cell bodies in the trigeminal ganglion and the sympathetic fibres have their cell bodies in the superior cervical ganglion. The parasympathetic fibres vary with regard to the site of their cell bodies.

THE SYMPATHETIC SYSTEM

The sympathetic fibres arise in the lateral horn of the grey matter of the spinal cord and leave the cord in the ventral nerve roots. They pass from the nerve to a sympathetic ganglion as a *white (myelinated) ramus communicans*; after synapse they leave the ganglion either to be distributed independently or to rejoin the spinal nerve as a *grey (non-myelinated) ramus communicans*.

The sympathetic trunk

There is one on each side and it extends the whole length of the vertebral column. Each consists of ganglia, united by intervening cords, and is placed partly in front and partly at the side of the vertebrae. Above, the trunk continues as a plexus into the cranium and below the trunks join together in front of the lower part of the sacrum in the *ganglion impar*. The ganglia are named according to the region they occupy—*cervical*, *thoracic*, *lumbar* and *sacral*. The cervical part has three ganglia, the thoracic eleven or twelve and the lumbar and sacral, four each.

The ganglia give off to the ventral ramus of the corresponding spinal nerve at least one grey postganglionic communicating branch; the ventral rami of the thoracic and first two lumbar nerves give off white preganglionic communicating branches to the ganglionated trunk. The interganglionic cords are composed of preganglionic fibres which after entering the trunk either (a) synapse at their own level, (b) ascend and synapse higher up, (c) descend and synapse lower down, or (d) leave the trunk and synapse in the collateral (usually aortic) ganglia (splanchnic nerves, p. 300).

CERVICAL PART OF SYMPATHETIC TRUNK

This lies behind the carotid sheath, adherent to the prevertebral fascia and has three ganglia.

(1) *Superior cervical ganglion*
Flattened and about 2.5 cm long; lies opposite the 2nd and 3rd cervical vertebrae. It has the following branches:

Communicating: To the ventral rami of 1st, 2nd, 3rd and 4th cervical nerves, the superior and inferior ganglia of the vagus, the hypoglossal nerve and the inferior ganglion of the glossopharyngeal.

Pharyngeal nerves and plexus: The pharyngeal nerves are given off from the front of the superior cervical ganglion and pass forwards and downwards; join with branches from the vagus and glossopharyngeal nerves to form the pharyngeal plexus which lies on the middle constrictor muscle.

Superior cardiac nerves: The right passes downwards behind the carotid sheath in front of the inferior thyroid artery and recurrent laryngeal nerve. It then goes either behind or in front of the subclavian artery and along the brachiocephalic artery to end in the deep cardiac plexus.

The left superior cardiac nerve has the same course as the right in the neck; in the thorax it passes along the left common carotid artery. It passes to the left of the arch of the aorta and joins the superficial cardiac plexus.

Branches to vessels: Filaments are given to the external carotid artery which continue along its branches.

Internal carotid nerve: Lies behind the internal carotid artery; enters the carotid canal and forms the carotid plexus on the lateral side of the internal carotid artery.

Before entering the cavernous sinus with the artery, communicating branches are given off to the tympanic branch of the glossopharyngeal nerve, the abducent nerve and trigeminal ganglion. Another branch is the deep petrosal nerve which joins the greater petrosal nerve to form the nerve of the pterygoid canal.

In the cavernous sinus the plexus gives off communicating branches to the oculomotor, trochlear and ophthalmic division of trigeminal nerve. It gives off the sympathetic branch to the ciliary ganglion; usually joins the sensory branch to the ganglion from the nasociliary of the ophthalmic (vasoconstrictor). The long ciliary nerves contain the sympathetic fibres which are motor to dilator pupillae (trigeminal fibres of long ciliary nerves are sensory to iris and ciliary body).

(2) *Middle cervical ganglion*
Usually lies on the inferior thyroid artery at the level of the 6th cervical vertebra. It has the following branches:

Communicating: To the ventral rami of the 5th and 6th cervical nerves.

Thyroid: To the thyroid gland, communicating with external and recurrent laryngeal nerves. Distributed along inferior thyroid artery to larynx and pharynx below level of vocal folds and trachea and oesophagus.

Middle cardiac nerve: On the right passes in front or behind subclavian artery to the front of the trachea and joins the deep cardiac plexus. On the left it lies between the left common carotid and left subclavian arteries and joins the deep cardiac plexus.

(3) *Inferior cervical ganglion*

Lies between the transverse process of the 7th cervical vertebra and the neck of the 1st rib, behind the vertebral artery, medial to the costocervical trunk. It has the following branches:

Communicating: To the ventral rami of the 7th and 8th cervical nerves.

Inferior cardiac nerve: Passes behind the subclavian artery, joins recurrent laryngeal and enters the deep cardiac plexus.

Branches to vessels: Branches are given to form a plexus round the vertebral artery.

Stellate ganglion

When present lies over neck of 1st rib. It is a fusion of inferior cervical with 1st thoracic ganglion.

Ansa subclavia

Formed of fibres which pass from middle cervical ganglion in front of subclavian artery and loop upwards behind artery to join the middle cervical ganglion. The main part of cervical sympathetic trunk passes into neck behind that vessel.

THORACIC PART OF SYMPATHETIC TRUNK

This lies by the side of the vertebrae on the necks of the ribs near their heads. The lowest two are on the bodies of the vertebrae. Bound down by parietal pleura anterior to intercostal neurovascular bundles. There are eleven or twelve ganglia and each has grey and white rami connecting it with the corresponding intercostal nerve.

Branches from the upper six ganglia are given off to the

thoracic aorta, vertebrae, ligaments, from the 3rd and 4th to
the oesophageal and pulmonary plexuses, and from the
2nd to the 5th to the deep cardiac plexus.

Branches from the lower six or seven ganglia include (a)
the *greater splanchnic nerve*; formed by the union of branches
from the 5th, 6th, 7th, 8th and 9th ganglia; passes medially
over the bodies of the vertebrae, perforates the crus of the
diaphragm and ends in the coeliac ganglion; (b) the *lesser
splanchnic nerve* from the 10th and 11th ganglia; runs with
the greater splanchnic nerve and ends in the coeliac ganglion; and (c) the *lowest splanchnic nerve* from the 12th
ganglion (if present); pierces the crus of the diaphragm and
ends in both the renal and the coeliac plexuses.

LUMBAR PART OF SYMPATHETIC TRUNK

This lies nearer the midline than the thoracic—on the
bodies of the vertebrae at the medial margin of psoas. The
sympathetic trunk enters the abdomen by passing behind
medial arcuate ligament and enters the pelvis behind the
common iliac artery.

Upper two lumbar ganglia receive white rami; all give
grey rami to corresponding lumbar nerves and in this way
to the branches of the lumbar plexus; rami pass behind
fibrous arches of psoas. The ganglia give branches to the
coeliac, aortic and superior hypogastric plexuses, and to the
plexuses round the blood vessels.

SACRAL PART OF SYMPATHETIC TRUNK

This lies medial to the anterior sacral foramina and is united
with the opposite trunk in front of the lower end of the
sacrum by a cord, in the middle of which there is the
ganglion impar.

Branches are given from the ganglia to sacral nerves and
thus to the branches of the sacral plexus. There are branches to the hypogastric plexuses.

The plexuses of the autonomic system

THE CARDIAC PLEXUSES

These are divided into two parts—*supeficial (ventral)* and *deep (dorsal)*.

The *superficial cardiac plexus* lies in the concavity of the arch of the aorta in front of the ligamentum arteriosum. It receives the left superior cardiac nerve of the sympathetic, the inferior cervical cardiac of the left vagus nerve, and branches from the deep plexus. The plexus gives branches to the anterior pulmonary plexus of the left side, and the right coronary plexus which accompanies the right coronary artery.

The *deep cardiac plexus* lies deep to the arch of the aorta on the bifurcation of the trachea. It consists of right and left parts which freely communicate. The right lies above the right branch of the pulmonary artery; the left lies on the left of the trachea, close to the ligamentum arteriosum.

It receives (a) all the cardiac branches from the cervical ganglia of the sympathetic, except the left superior nerve, and (b) all the cardiac branches of the vagus and recurrent laryngeal nerves, except the inferior cervical cardiac of the left vagus.

Branches from the right side join the superficial cardiac plexus to form the right coronary plexus, and others are distributed to the right atrium. Branches from the left side end mainly in the left coronary plexus which accompanies the left coronary artery, and in the superficial cardiac plexus.

THE PULMONARY PLEXUSES

These are divided into two parts—a large posterior and smaller anterior, anterior and posterior to the hilum of the lung. The vagus (parasympathetic) goes mainly to the posterior plexus and gives a small branch to the anterior. The sympathetic fibres come from the 2nd to the 5th thoracic ganglia. Both contain motor and sensory fibres.

THE OESOPHAGEAL PLEXUS

This is related to the lower half of the thoracic oesophagus. The anterior part receives mainly the left vagus and the posterior the right vagus. The plexus also receives sympathetic fibres from the 3rd and 4th thoracic ganglia. The two parts continue into the abdomen with the oesophagus as the *posterior vagal trunk (posterior gastric nerve)* which divides into a small gastric and large coeliac branch, and the *anterior vagal trunk (anterior gastric nerve)* which goes mainly to the stomach but also to the liver.

THE COELIAC PLEXUS

This is the largest prevertebral plexus. It lies in front of the aorta and crura of the diaphragm behind the pancreas and inferior vena cava. It surrounds the origin of the coeliac trunk and extends laterally to the suprarenal glands. It receives the greater and lesser splanchnic nerves and part of the right vagus. It contains several ganglia and gives off branches which accompany the blood vessels to the viscera and form secondary plexuses on these arteries.

The *coeliac ganglia*, one in each half of the coeliac plexus, lie on the medial side of the suprarenal glands, the right lying behind the inferior vena cava. The greater splanchnic nerve enters its upper end.

The remaining plexuses in the abdomen may be regarded as extensions of the coeliac.

The *phrenic plexus* accompanies the arteries to the diaphragm. On the right side near the suprarenal gland is the *phrenic ganglion*.

The *suprarenal plexus* is derived from the coeliac plexus and the lateral part of the coeliac ganglion. It is joined by branches of the greater splanchnic nerve. The nerve fibres in the suprarenal gland are preganglionic although many are non-myelinated.

The *renal plexus* is derived from the coeliac ganglion and partly from the coeliac and aortic plexuses, and receives the lowest splanchnic nerve. It lies along the renal artery, and contains numerous small ganglia. The fibres are mainly vasomotor.

The *testicular (ovarian) plexus* comes from the renal and aortic plexuses and lower down from the hypogastric plexuses.

The *left gastric plexus* accompanies left gastric artery along the lesser curvature of the stomach and communicates with the vagus nerves.

The *hepatic plexus* accompanies hepatic artery into the substance of the liver. It communicates with the left vagus and the right suprarenal plexus, and gives off the cystic, pyloric, right gastro-epiploic and pancreaticoduodenal plexuses.

The *splenic plexus* accompanies splenic artery to the spleen, and is joined by the right vagus. It gives off the left gastro-epiploic and pancreatic plexuses.

The *superior mesenteric plexus* accompanies the superior mesenteric artery.

The *abdominal aortic (intermesenteric) plexus* lies on the anterior surface of the abdominal aorta. It gives off the *inferior mesenteric* and part of the testicular (ovarian) plexuses. It continues inferiorly as the *superior hypogastric plexus*.

The *superior hypogastric plexus (presacral nerve)* lies on the sacral promontory between the two common iliac arteries; it is formed by the termination of the aortic plexus, together with branches from the 3rd and 4th lumbar ganglia. It divides below into *right* and *left hypogastric nerves*. The hypogastric plexus gives branches to the ureteric, gonadal and common iliac plexuses. It may contain parasympathetic fibres which ascend from the inferior hypogastric plexus.

The *right* and *left hypogastric nerves* run downwards in the pelvis medial to the internal iliac artery on each side, to become the *inferior hypogastric plexuses* which lie on each side of the rectum. The plexus extends forwards in the male to the sides of the prostate and bladder and in the female to the sides of the uterus, vagina and bladder and laterally into the base of the broad ligament.

The plexus contains sympathetic (mainly postganglionic) fibres from the superior hypogastric plexus and sacral ganglia and parasympathetic (preganglionic) from 2nd, 3rd and 4th sacral spinal cord segments as well as ganglion cells.

The following plexuses are derived from the inferior hypogatric plexuses:

The *rectal plexus* to the rectum and anal canal.

The *vesical plexus* to the bladder, with secondary plexuses in the male to the ductus deferens and to the seminal vesicles.

The *prostatic plexus* to the prostate gland, giving off the cavernous nerves of the penis.

The *vaginal plexus* to the vagina.

The *uterine plexus* accompanying the uterine artery to the uterus.

Appendix to chapter 7

The arrangement and distribution of the autonomic nervous system

	Origin	Nerves carrying preganglionic fibres	Cell station and synapse	Distribution	Effect of stimulation
Cranial parasympathetic	Midbrain	Oculomotor	Ciliary ganglion	Sphincter pupillae and ciliary muscle	Contraction of the pupil and accommodation
	Hindbrain	Facial	Submandibular and pterygopalatine ganglia	Submandibular, sublingual and lacrimal glands	Increased flow of saliva and tears
		Glossopharyngeal	Otic ganglion	Parotid gland	Increased flow of saliva
		Vagus	Mostly in the viscera supplied	Heart and lungs	Slowing of heart beat; contraction of smooth muscle and secretion of mucus in lungs
				Alimentary canal, liver, gall bladder and pancreas	Increased peristalsis and relaxation of sphincters

continued

Appendix to chapter 7 (*continued*)

	Origin	Nerves carrying preganglionic fibres	Cell station and synapse	Distribution	Effect of stimulation
Sympathetic	Thoracolumbar part of spinal cord	1st thoracic to 2nd lumbar nerves	Ganglia of the sympathetic trunk; also ganglia in the cardiac, pulmonary, oesophageal, coeliac, aortic, mesenteric and hypogastric plexuses	Dilator pupillae (T.1)	Dilatation of the pupil
				Sweat glands, arrectores pilorum	Secretion of sweat, erection of hairs
				Blood vessels	Vasoconstriction (vasodilatation of blood vessels of heart and skeletal muscle)
				Lungs	Relaxation of smooth muscle in lungs
				Heart	Acceleration of the heart beat
				Alimentary canal and its associated glands of secretion	Inhibition of peristalsis Contraction of sphincters of bladder and rectum
Sacral parasympathetic	Lowest part of spinal cord	2nd, 3rd and 4th sacral nerves	Mostly in the viscera supplied	Bladder	Emptying of bladder and rectum, their sphincters being at the same time relaxed. Erection of external genitalia
				Rectum	
				Genitalia (i.e. derivatives of cloaca)	

8
The Ear and Eye

THE EAR

The ear is composed of three parts—external, middle and internal.

THE EXTERNAL EAR

The external ear consists of the *auricle* and *external acoustic (auditory) meatus.*

The *auricle (pinna)* consists of elastic cartilage covered with skin, forms the outer third of the meatus and is attached to its inner bony part. It has numerous ridges and depressions. The projection in front of the meatal opening is called the *tragus* and the hollow behind, the *concha*. The outer edge forms the *helix* and the varying-sized downward projection, the *lobule* (fibrofatty tissue, not cartilage).

The *external acoustic meatus* extends from the concha to the tympanic membrane; it is 3 cm long. It is arched slightly upwards, and is directed medially and slightly forwards; because of its direction the canal is straightened by pulling the auricle upwards and backwards. The canal is formed partly of cartilage and partly of bone. The lateral or cartilaginous part is continuous with the auricle, and is about 1 cm long. The cartilage does not form a complete tube and is deficient in its upper and back part; the interval is filled by fibrous tissue. At the medial end of the bony part there is a groove round the sides and floor for the attachment of the tympanic membrane. The external meatus is lined by skin in which are hairs and ceruminous glands secreting wax.

THE MIDDLE EAR (Fig. 64)

The middle ear is contained in the temporal bone. It communicates with the nasopharynx by the auditory (Eustachian) tube and is traversed by a chain of bones which

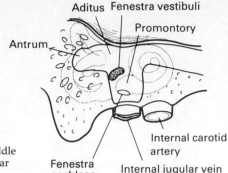

Fig. 64 View of medial wall of middle ear with internal ear superimposed.

transmit vibrations from the tympanic membrane to the internal ear.

TYMPANIC CAVITY

It is 15 mm high, 15 mm from front to back and 5 mm from side to side; bounded laterally by the meatus and tympanic membrane, medially by the lateral wall of the internal ear; its upper part (above the level of the tympanic membrane) is called the *epitympanic recess* and communicates posteriorly with the *mastoid (tympanic) antrum* by the *aditus*.

The *roof* is formed by a thin plate of bone (*tegmen tympani*) separating the middle ear from the middle fossa of the skull and the meninges and temporal lobe of the brain.

The *floor* is formed by the roof of the jugular fossa and is related to the internal jugular vein and internal carotid artery.

The *lateral wall* is formed by the tympanic membrane and the bone around it. Inferiorly the following openings are seen: (a) the *squamotympanic fissure*, through which the anterior ligament of the malleus and anterior tympanic branch of maxillary artery pass, (b) the *posterior canaliculus* from which the chorda tympani, a branch of the facial nerve emerges, and (c) the *anterior canaliculus* through which the chorda tympani passes to join the lingual nerve.

The medial wall has the following features: The *fenestra vestibuli (foramen ovale)* which leads into the vestibule and is closed by a ligament and the base of the stapes; the *promontory* below and in front of the fenestra vestibuli, formed by the projection of the first turn of the cochlea; the *fenestra cochleae (foramen rotundum)* behind the promontory. It is closed by a membrane, the *secondary tympanic membrane* which covers an aperture in the bone leading to the scala tympani of the cochlea. The prominence of the *facial canal*, for the facial nerve, runs backwards along the upper part of the medial wall and then turns downwards medial to the aditus.

The *posterior wall* has superiorly the large aperture (*aditus*) to the mastoid antrum below which is the *pyramid* just behind the fenestra vestibuli; it contains the *stapedius muscle*. Its tendon projects through the apex.

The *anterior wall* has two openings: the *canal for the tensor tympani*, just anterior to the fenestra vestibuli, is superior to the *auditory (Eustachian) tube* which leads into the pharynx. Part of the tube is elastic cartilage and part bone. The medial or cartilaginous part is trumpet-shaped, and terminates in an oval opening on the lateral wall of the nasopharynx. The osseous portion lies along the angle of union of the squamous and petrous parts of the temporal bone and is about 1.5 cm long. Below this a thin plate of bone separates the tympanum from the carotid canal. The *processus cochleariformis* is a process of bone lying between and separating the canal for the tensor tympani and the auditory tube.

The *tympanic membrane* separates the external and middle ears. It is inserted into the groove in the osseous portion of the external meatus, is placed obliquely downwards and medially across the opening and forms with the floor of the meatus an angle of 55°. It bulges into the middle ear so that its lateral surface is concave. Attached to its medial surface is the handle of the malleus. The upper part of the membrane is thin and loose, and is known as the *pars flaccida*.

Auditory Ossicles

(1) *Malleus* (hammer): Consists of a *head* which has on the upper part of its posterior surface a facet for articulation with the body of the incus; a *neck*, a constriction below the

head; a *handle*, a long tapering process passing downwards and backwards attached laterally to the tympanic membrane; an *anterior process*, a slender spicule passing from the neck downwards and forwards towards the petrotympanic fissure; a *lateral process* which arises from the root of the handle and projects laterally to be attached to the tympanic membrane by the anterior and posterior malleolar folds which bound the flaccid part of the membrane.

(2) *Incus* (anvil): Consists of a *body*, articulating in front by a saddle-shaped facet with the head of the malleus; a *short process*, attached to the margin of the aditus on the posterior wall; a *long process*, passing downwards behind and parallel to the handle of the malleus with its tip projecting medially and articulating with the head of the stapes.

(3) *Stapes* (stirrup): Consists of a *head* facing laterally and articulating with the incus; a *base* fixed to the membrane closing the fenestra vestibuli; *limbs (crura)* arising from a constricted part, the *neck*, and passing medially to the extremities of the base. The anterior crus is shorter and straighter than the posterior.

Ligaments unite the chain of bones to the adjacent walls of the cavity; *anterior ligament of malleus* between root of anterior process and petrotympanic fissure; *lateral ligament of malleus* between lateral process and posterior malleolar fold; *superior ligament of malleus* between head of the malleus and roof; *posterior ligament of incus* between short process and posterior wall; *annular ligament* of stapes between base of stapes and edge of fenestra vestibuli.

In addition *capsular ligaments* of elastic tissue surround the synovial joints between the malleus and incus, and incus and stapes.

The muscles

The *tensor tympani* arises from the cartilage of the auditory tube and from the bony canal in which it lies. The tendon turns round the end of the processus cochleariformis and is inserted into the medial border of the handle of the malleus near its root. It is supplied via otic ganglion by mandibular division of trigeminal nerve.

The *stapedius* is lodged in the pyramid, from the apex of which its tendon issues to be attached to the posterior part of the neck of the stapes. It is supplied by facial nerve.

Action of both is to damp down over-vibration of membrane and ossicles.

Mastoid (Tympanic) Antrum

This is a large recess behind and above the tympanic cavity of the middle ear with which it is connected by an opening (*aditus*) on the upper part of the posterior tympanic wall; it is developed with the cavity and is lined by a continuation of its mucous membrane. Into it open the *mastoid air cells*. It lies deep to the *suprameatal triangle* 15 mm from the surface in the adult (5 mm in an infant).

THE INTERNAL EAR

Within the internal ear are the fibres of the vestibulocochlear (eighth cranial) nerve. The internal ear is divided into *bony* and *membranous labyrinths* (Fig. 65), the former enclosing the latter. The *endolymph* is the fluid within the membranous labyrinth and the *perilymph* the fluid between the membranous and osseous labyrinths.

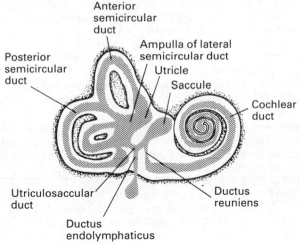

Anterior
semicircular
duct

Ampulla of lateral
semicircular duct

Posterior
semicircular
duct

Utricle

Saccule

Cochlear
duct

Utriculosaccular
duct

Ductus
reuniens

Ductus
endolymphaticus

Fig. 65 Membranous and bony labyrinths.

The osseous labyrinth

This consists of the *vestibule, semicircular canals* and *cochlea*. The *vestibule* is the central part. Its lateral wall corresponds with the medial wall of the middle ear and in it is the fenestra vestibuli, closed by the base of the stapes; on its medial wall is a depression perforated by several holes for the fibres of the eighth nerve; behind is a ridge, the *vestibular crest*; and still further back is the medial opening of the *aqueduct of the vestibule*, for transmission of the *endolymphatic duct*. On the roof is a depression, the *elliptical recess*. Posteriorly are the five openings of the semicircular canals; anteriorly is an opening which leads to the scala vestibuli of the cochlea.

The *semicircular canals*, three arched osseous canals, are placed above and behind the vestibule and open into that chamber by five rounded apertures (two adjacent canals have a common opening). Each canal forms about two-thirds of a circle, and has at one end a dilated part, the *ampulla*.

The *anterior (superior) canal* is vertical, and placed transverse to the long axis of the petrous temporal bone (that is, at about 45° to the sagittal plane). The medial extremity joins the opening of the posterior canal.

The *posterior canal* is vertical in a plane at right angles to the former. Its upper end is joined to the lower opening of the superior canal.

The *lateral canal* is the smallest of the three, and is almost horizontal. It is just above the level of the fenestra vestibuli.

The *cochlea* (the organ of hearing) (Fig. 66) is shaped like a snail's shell lying on its side so that the base is turned to the internal meatus, and the apex is opposite the canal for the tensor tympani. The cochlea consists of a tapering spiral canal of 2½ turns, around a bony axis or *modiolus*; the canal is divided into three compartments by partitions of bone (*spiral lamina*) and membranes. The first turn of the canal bulges into the middle ear and forms the promontory. The osseus spiral lamina ends at the apex of the cochlea and between it and the modiolus is a small opening, the *helicotrema*, by which the upper and lower compartments communicate.

The modiolus is pierced by small canals for the fibres of the cochlear nerve. A small canal, which winds round the modiolus contains the *spiral ganglion* of the cochlear nerve. The peripheral processes of the ganglion cells go to the organ of Corti and the central processes form the cochlear nerve.

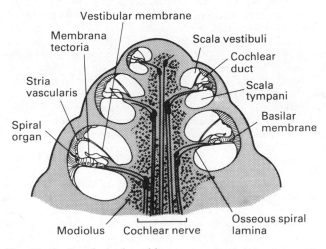

Vestibular membrane

Membrana tectoria

Scala vestibuli

Cochlear duct

Stria vascularis

Scala tympani

Spiral organ

Basilar membrane

Modiolus   Cochlear nerve

Osseous spiral lamina

Fig. 66   Section through cochlea.

The three compartments in the cochlea are the *scala tympani* (lower), the *cochlear duct* (middle) and *scala vestibuli* (upper). The scala tympani commences at the fenestra cochleae. Near the fenestra cochleae is the opening of the *aqueduct of the cochlea* which conveys perilymph to an aperture in the glossopharyngeal notch of the petrous bone in the jugular foramen, where it communicates with the cerebrospinal fluid.

The *scala vestibuli* commences at the fenestra vestibuli and communicates at the apex of the modiolus with the scala tympani by the helicotrema. The scalae contain perilymph.

The membranous labyrinth

This consists of sacs containing endolymph. The fibres of the vestibulocochlear nerve are distributed on its wall. The membranous sac in the bony cochlea lies between the scala tympani and the scala vestibuli and forms the *duct of the cochlea*.

In the bony vestibule there are two sacs, the *utricle* and the *saccule*. The utricle is larger than the saccule, and is situated in the posterior and upper part of the vestibule. The membranous semicircular ducts open into its posterior part. The vestibular nerves enter the anterior part. The interior has a thickened area called the *macula* on which there is a small mass of calcareous grains known as *otoliths*. Inferiorly there is a canal, which extends along the aqueduct of the vestibule, the *endolymphatic duct*, which ends in a dilated pouch, the *endolymphatic sac*.

The saccule is smaller and rounder than the utricle. Like the utricle, it contains a macula and otoliths. Below there is a small canal, *ductus reuniens*, which connects it with the duct of the cochlea.

The *semicircular ducts* are about one-third the size of the canals, except at the ampullae, where each dilates nearly to fill the bony canal. Each duct is free on its concave aspect and the convexity is fixed to the osseous canal. On the surface of the ampulla where it is attached to the bony canal there is a transverse projection, the *ampullary crest*, in which some filaments of the vestibular nerve end.

The *membranous cochlea (cochlear duct)* separates the scalae vestibuli from the scala tympani. The duct contains the *spiral organ* to which the cochlear nerve is distributed and is separated from the scala vestibuli by the *vestibular membrane* which passes from the spiral lamina upwards and laterally to the wall of the bony cochlea. The duct is bounded below by the *basilar membrane* to which the spiral organ is attached and is connected to the saccule by the *ductus reuniens*. Above, it terminates in a blind cone-shaped extremity, partly bounding the helicotrema.

The *vestibulocochlear (auditory) nerve* enters the internal acoustic meatus and divides into two branches which enter foramina at the bottom of the meatus and are distributed to the vestibule and cochlea.

The *vestibular branch* supplies the maculae of the saccule and utricle and the cristae of the ampullae of the semicircular ducts. The *vestibular ganglion* lies deep in the internal acoustic meatus.

The *cochlear branch* is distributed to the cochlea. Its fibres perforate a number of foramina in the centre of the base of the cochlea. These foramina lead to small canals which at first pass through the modiolus and then radiate laterally between the bony layers of the spiral lamina to the spiral organ. Bipolar neurons in the modiolus at the base of the spiral lamina form the *spiral ganglion*.

LACRIMAL APPARATUS

The *eyebrows* are arched eminences, one over each orbit. They consist of thickened skin and muscle, surmounted by hairs.

The *eyelids* are two movable folds, an upper and a lower, which by their closure protect the eye. Blinking spreads a film of tears on the cornea and pumps away excessive tears. When the eyelids are open the angles of junction of the upper and lower lids are called the *lateral* and *medial angles (canthi)*. At the medial angle there is a small triangular area, the *lacus lacrimalis*, in which there is a pink mass of connective tissue containing sebaceous glands, the *lacrimal caruncle*. This is separated from the eyeball by a vertical fold of conjunctiva, the *plica semilunaris*, a rudimentary third eyelid. At the lateral edge of the caruncle, on each lid, is the *lacrimal papilla*, which is pierced by the *punctum lacrimale*, the external opening of the *lacrimal canaliculus*.

From without inwards the eyelid consists of skin, areolar tissue, orbicularis oculi muscle, tarsus and orbital fascia, tarsal glands and conjunctiva; the upper lid also contains the aponeurosis of the levator palpebrae superioris which is attached to the upper margin of the tarsal plate (tarsus).

The tarsi or *tarsal plates* are laminae of condensed connective tissue found in each lid; the superior, the larger, is half oval in shape, the lower a narrow oblong strip. In their substance are lodged the *tarsal glands*. Apart from the free edge in the eyelid each tarsus is continuous with the membranous sheet known as the *orbital fascia*. Medially it is attached to the *medial palpebral ligament*. The superior tarsus receives the main insertion of the levator palpebrae superioris from above.

The *palpebral fascia* forms an incomplete diaphragm for the anterior orifice of the orbit; peripherally it is attached to orbital margin and centrally to the edge of the tarsus.

The *conjunctiva* is modified skin and forms the most posterior layer of both eyelids, at the free edges of which it joins the outer skin. It is firmly attached to the deep surface of each tarsal plate (prevents wrinkling); beyond this it is reflected on to the eyeball; the lines of reflection are known as the *fornices* of which the superior is the deeper. Over the eyeball, the conjunctiva is loosely connected to the sclera. Above and laterally the ducts of the lacrimal gland open into the conjunctival sac.

The *lacrimal gland* occupies a depression in the superolateral angle of the orbit; the anterior margin is connected to the back part of the upper eyelid. The ducts (12 or 14) open in a row into the superior conjunctival fornix.

Each lacrimal canaliculus commences at the punctum lacrimale on the free edge of the lid near the caruncle. The canaliculus arches medially in the free edge of the lid to open into the *lacrimal sac* which lies in a groove formed by the lacrimal bone and the frontal process of the maxilla, behind the medial palpebral ligament and in front of the lacrimal part of the orbicularis oculi muscle; it is the dilated upper end of the *nasolacrimal duct* which is formed by the lacrimal bone, maxilla and inferior nasal concha. It leads from the lacrimal sac to the inferior meatus of the nose where it opens; the aperture is partly guarded by the lacrimal fold formed by the mucous membrane. It is about 7 mm long and is directed downwards and slightly laterally and backwards.

THE EYEBALL

The eyeball is contained within the orbit; it is spherical but anteriorly the cornea is the segment of a smaller sphere. It consists of three coats within which there are three refractive media:

Outer coat, fibrous—*sclera* and *cornea*.

Middle coat, vascular, pigmented—*choroid*, *ciliary body* and *iris*.

Inner coat, nervous—*retina*.

The refractive media are (a) *aqueous humour*, (b) *vitreous body*, (c) *lens*.

The *fascial sheath of the eyeball (capsule of Tenon)* covers the posterior five-eighths of the eyeball and is continuous posteriorly with the sclera at the exit of the optic nerve. Anteriorly it is connected to the sclerocorneal junction by loose tissue. The fascial sheath is pierced by the tendons of the extrinsic muscles of the eyeball and extends along the tendons for a short distance. It is connected with the eyeball only by delicate connective tissue; the interval forms a potential space like a bursa and facilitates movements of the eyeball.

The *sclera* is opaque and fibrous, and occupies the posterior five-sixths of the eyeball; it is continuous in front with the cornea at the *sclerocorneal junction*. The outer surface is white and smooth and has attached to it the recti and oblique muscles.

The inner surface is of a light brown colour due to a lining of pigmented connective tissue, the *lamina fusca*, which is connected by fine filaments to the choroid coat. Between the sclera and choroid is a space containing branches of the ciliary vessels and nerves.

The optic nerve passes through small openings in the posterior part of the sclera, about 3 mm medial to the axis of the eyeball; this area is called the *lamina cribrosa sclerae*. The outer sheath of the nerve (dura mater) blends with the sclera (Fig. 67).

The sclera is thickest at its posterior part, gradually thinning until about 6 mm from the cornea, where it thickens again.

In the sclera close to its junction with the cornea there is a small circumferential canal called the *sinus venosus sclerae*.

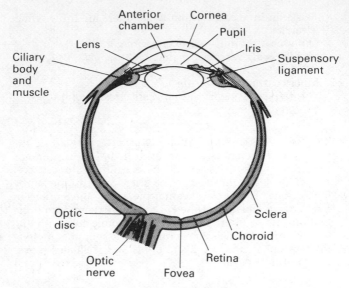

Fig. 67 Horizontal section through eyeball.

The blood vessels of the sclera are few in number, but near its junction with the cornea there is a vascular zone derived from the anterior ciliary branches of the ophthalmic artery.

The *cornea* is the anterior transparent part of the outer coat of the eyeball, occupying about one-sixth of the circumference of the globe. Being the segment of a smaller sphere it projects forwards beyond the curvature of the sclera. The posterior surface is concave, and projects further than the anterior convex surface so that it is overlapped by the edge of the sclera; this surface forms the anterior boundary of the anterior chamber of the eye which contains the aqueous humour (Figs. 67 and 68).

At the circumference of the cornea some of the fibres which form its stroma are continued backwards and outwards into the sclera and iris; those going to the iris are called the *pectinate ligament of the iris*; they form an annular

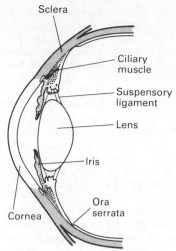

Fig. 68   Section through anterior part of eyeball.

meshwork enclosing a series of spaces (*spaces of iridocorneal angle*) which communicate with the anterior chamber.

The *vascular coat* is situated between the sclera and the retina. It is continued anteriorly into the *ciliary body*, forming a number of projections (*ciliary processes*) which fold inwards and are arranged in a circle behind the iris.

The *choroid* part of the vascular coat is thickest behind where it is pierced by the optic nerve. Externally it is connected to the sclera by loose connective tissue (*suprachoroid lamina*) traversed by vessels and nerves. Internally it is lined by the pigmented cells of the retina.

The choroid consists of blood vessels connected together by loose connective tissue and contains large branched and pigmented cells. It is made up of two layers, an outer and an inner. The outer layer (*vascular lamina*) contains the larger branches of the vessels. The arteries, the short posterior ciliary, pierce the sclera close to the optic nerve, pass forwards and bend inwards to end in the inner layer. The veins are external to the arteries and join together into four or five principal trunks which pierce the sclera midway between the cornea and the optic nerve.

The inner coat (*choriocapillary lamina*) is formed by the capillary endings of the vessels of the outer coat; they pass forwards to about 3 mm from the cornea, and join the capillaries of the ciliary processes.

The *ciliary body* consists of the *ciliary processes* and the *ciliary muscle*.

The *ciliary processes* have the same structure as the rest of the choroid. There are about seventy and they are placed in corresponding depressions upon the surface of the vitreous body, and upon the ciliary zonule. The blood vessels are derived from the anterior ciliary branches of the arteries to the recti muscles.

The *ciliary muscle* consists of two sets of involuntary muscular fibres, meridional and circular. It is supplied by the parasympathetic part of the oculomotor nerve. The meridional fibres arise from the sclera close to its junction with the cornea and are inserted into the choroid opposite the ciliary processes. The circular form a zone of circular fibres internal to the meridional at the base of the ciliary processes.

Contraction of the ciliary muscle produces relaxation of the suspensory ligament of the lens which is attached to the base of the ciliary processes. This results in the lens becoming more biconvex and allows focussing on near objects (*accommodation reflex*).

The *iris* is the coloured membrane suspended in the aqueous humour behind the cornea and in front of the lens. In the centre is a circular aperture, the *pupil* (Fig. 67).

It is continuous at its circumference with the ciliary body. Anterior to the ciliary body it is attached to the cornea by the pectinate ligament of the iris.

The anterior surface is coloured and marked by wavy lines converging towards the free edge of the pupil.

The posterior surface is darkly pigmented (*uvea*) and marked with folds prolonged from the ciliary processes.

The framework of the iris is a delicate stroma of connective tissue, containing blood vessels, nerves, pigment cells and two groups of involuntary muscular fibres. The *sphincter of the pupil* is a narrow band of fibres placed close to the pupil. It is supplied by the parasympathetic part of the oculomotor nerve. The *dilator of the pupil* commences at the outer margin of the iris and its radial fibres converge towards the pupil. It has a sympathetic nerve supply (p. 298).

The *blood vessels* of the iris come from the two long posterior ciliary and the anterior ciliary arteries; the former pierce the sclera close to the optic nerve, and pass forwards in the space between the lamina fusca of the sclera and the suprachoroid lamina of the choroid, divide into two branches and enter the iris. They anastomose with the corresponding vessels of the opposite side, and with vessels from the vascular zone of the sclera which are formed by the anterior ciliary arteries. These form the *greater arterial circle of the iris*. Small branches from this circle converge towards the pupil, and freely anastomose with one another, forming the *lesser arterial circle*.

The *veins* follow the same arrangement as the arteries and drain into the veins of the choroid coat.

The *ciliary nerves* of the choroid and iris are about fifteen in number. They come from the ciliary ganglion and the nasociliary branch of the ophthalmic nerve; they follow very closely the course of the blood vessels and on reaching the ciliary body, form a plexus which supplies twigs to the ciliary muscle, iris and cornea.

The *retina* is the expanded termination of the optic nerve, and forms the innermost coat of the eye. It reaches forwards nearly as far as the ciliary processes, where it ends in a saw-edged border, the *ora serrata*; from this border there is prolonged a thin layer as far as the ciliary processes, which blends with the iris. This prolongation contains no nerve fibres, and is called the *ciliary part of retina* (Fig. 68).

The outer surface is covered with pigment cells (formerly described as part of the choroid).

The inner surface which can be examined with an ophthalmoscope shows (a) the *macula lutea*, or yellow spot, situated in the axis of the globe. It has a depression in it (*fovea centralis*); and (b) the *optic disc* (about 2 mm to the medial side of the yellow spot) towards which the optic nerve fibres converge and where they leave the eyeball as the *optic nerve*. The branches of the central artery and vein of the retina can be seen radiating from the optic disc.

The *central artery of the retina* travels in the optic nerve and reaches the inner surface of the retina at the disc. It divides into two branches, a superior and inferior, and each of these divide into a lateral or temporal division, and a medial or nasal.

The lateral branches give small branches which end in capillaries round the fovea. The rest of the branches are distributed as capillaries, to the retina as far as the ora serrata, but the smaller branches do not anastomose with one another or with any other vessels. The veins follow the same distribution as the arteries.

The *vitreous body* is a soft gelatinous substance occupying about four-fifths of the eyeball. It supports the retina behind, and is hollowed out in front for the lens.

Between the retina and the vitreous, and enclosing the latter except in front, is a thin capsule, the *hyaloid membrane*. Anteriorly it passes forwards to the margin of the lens. It becomes stronger in this part, and is called the *ciliary zonule (suspensory ligament)*.

The zonule commences near the ciliary processes and passes forwards to the front of the lens where it is attached to the lens capsule. Some fibres are attached to the extreme edge of the lens and others become continuous with the posterior part of the capsule. The interstices between these fibres are occupied by fluid, but after death they may be distended with air, so that the lens appears to be encircled by a canal (*zonular space*).

Extending forwards from the optic disc through the vitreous, as far as the capsule of the lens, is the *hyaloid canal*, the remains of a passage in the fetus for a branch from the central artery of the retina.

The *lens* is a transparent biconvex body enclosed in a transparent membrane, the *lens capsule*. It is in contact anteriorly with the iris; posteriorly it rests in a depression in the vitreous body, and it is surrounded by the suspensory ligament, connecting it to the ciliary processes and ciliary muscle. It is about 8 mm in diameter, and about 5 mm thick.

The capsule of lens is the structureless membrane enclosing the lens, thick in front near its circumference, where it is strengthened by the fibres of the zonule, but very thin posteriorly.

The *aqueous humour* occupies the space between the anterior surface of the lens capsule and the posterior surface of the cornea. The iris divides the chamber into two parts, known as the *anterior* and *posterior chambers*.

The anterior chamber is bounded in front by the cornea and behind by the iris (Fig. 67).

The posterior chamber is the triangular interval at the circumference of the lens bounded anteriorly by the ciliary processes and the iris and posteriorly by the suspensory ligament.

THE EXTRINSIC MUSCLES OF THE EYE

These are the four *rectus* and two *oblique muscles*. With the exception of the inferior oblique they all arise from a circumscribed area at the back of the orbit, pass forwards to surround the eyeball and form the cone of muscles. The four recti arise from a fibrous band, the *common tendinous ring*, surrounding the optic foramen and crossing the medial part of the superior orbital fissure. Each passes forwards and is inserted into the eyeball behind the corneoscleral junction. The *recti* are *superior* and *inferior* (rotating the eyeball upwards and downwards respectively about a transverse axis), and *lateral* and *medial* (rotating the eyeball outwards and inwards respectively about a vertical axis). The *superior oblique* arises from the upper and medial part of the common tendinous ring, passes forwards along the junction of the roof and medial wall of orbit, and then through the pulley (trochlea) which deflects its tendon in a hairpin bend backwards and laterally inferior to the superior rectus. It is inserted into the upper and lateral aspect of the globe behind its equator. The *inferior oblique* arises from the medial part of the front of the floor of the orbit, passes

laterally inferior to the inferior rectus and between the
lateral rectus and the eyeball, to be inserted into the lower
and lateral quadrant of the globe behind its equator.

Nerve supply: Oculomotor to all except lateral rectus
supplied by abducent, and superior oblique supplied by
trochlear.

Actions: The lateral and medial recti act in a horizontal
plane, but the superior and inferior recti pull the front of
the eyeball medially as well as upwards or downwards.
This is corrected thus:

Superior rectus: up and in
Inferior oblique: up and out   } Result: vertically up

Inferior rectus: down and in
Superior oblique: down and out   } Result: vertically down

The eyeball can to some extent also rotate about a
longitudinal axis passing through its anterior and posterior
poles (*torsion*). It is said that almost any movement of the
eyeball involves at least three muscles.

9
The Alimentary System

This extends from the mouth to the anus and includes the liver and pancreas.

THE MOUTH

This, the first part of the alimentary tract, opens on to the exterior through the aperture between the lips. The *lips* are two fleshy folds consisting mainly of muscles (p. 50). The external skin meets the internal mucous membrane at the *red margin*. Both skin and membrane are firmly bound down to the underlying tissue. The lips meet laterally at the *angles* of the mouth and a thickened nodule can be felt at the angle where the fibres of the various muscles meet. The vertical groove on the upper lip in the midline is called the *philtrum*.

The mouth is divided into an outer part, the *vestibule*, which is between the lips and cheeks and the teeth and gums, and the *oral cavity*. In the vestibule there is a vertical midline upper and lower mucosal fold (*frenulum*) between the lip and the gum. The parotid duct opens into the vestibule on a papilla inside the cheek opposite the 2nd upper molar tooth. There are many mucous glands in the mucous membrane of the lips and cheeks.

The *oral cavity* is bounded anteriorly and at the sides by the teeth and gums. It is roofed by the palate and the tongue is in the floor. In front of and below the tongue, the floor is formed by the mylohyoid muscles. Between the tongue and floor there is the *sublingual papilla* on each side of the *lingual frenulum*; the submandibular duct opens on it. Lateral to the papilla is *sublingual fold* overlying the sublingual salivary gland, the ducts of which open on the fold or into the submandibular duct.

THE TEETH

Deciduous (milk) teeth
These are five in each quadrant—two incisors, one canine and two molars. The dates in *months* of their eruption:

Molar 2	Molar 1	Canine	Incisor 2	Incisor 1
24	12	18	9	6

Permanent teeth
These are eight in each quadrant—two incisors, one canine, two premolars and three molars. The dates in *years* of their eruption:

Molar 3	Molar 2	Molar 1	Premolar 2	Premolar 1
18	12	6	10	9
	Canine	Incisor 2	Incisor 1	
	11	8	7	

Structure of a tooth
Pulp cavity of tooth surrounded by *dentine*, foramen in tip of root (*apical foramen*) transmits vessels, lymphatics and nerve filaments of pulp. Dentine is capped with *enamel* on visible part of tooth (*crown*) and covered with *cementum* on root. Teeth held in sockets by *periodontium* (*periodontal membrane*) which is vascular periosteum uniting bone of socket to cementum on root.

Incisors, canines and premolars have one root, upper molars three roots (two buccal, one palatal), lower molars two roots (one anterior, one posterior).

THE TONGUE

The tongue occupies the floor of the mouth; its *base* is connected with the hyoid bone, the epiglottis, the palatoglossal arches and the pharynx; along its inferior surface the genioglossus runs from base to tip, connecting it to the mandible and hyoid bone.

The mucous membrane on the inferior surface is smooth, forming in front a median fold, the *lingual frenulum*; on the sides it is continuous with the mucous membrane of the mouth. *Dorsum* of tongue (upper surface) lies partly in mouth (anterior two-thirds) and partly in oropharynx (posterior one-third). *Vallate papillae* mark the junction. Anterior two-thirds of dorsum covered with *papillae*. A shallow Λ-shaped *terminal sulcus* leads back in midline to *foramen caecum*. Posteriorly the epiglottis is connected to the tongue by three *glosso-epiglottic* folds, a *median* and two *lateral*, enclosing the *valleculae*. The *filiform papillae* (threadlike) are numerous and are arranged in rows parallel to the vallate, but towards the tip of the tongue their direction becomes more transverse. They are keratinized at their tips. The *fungiform papillae* are found principally at the apex and on the sides. They are not keratinized.

The *vallate papillae* (seven to ten) form a row on each side at the back of the tongue in front of the sulcus terminalis, and meet in front of the foramen caecum.

The fungiform and vallate papillae have taste buds. Those of the former are innervated by the chorda tympani, those of the latter by the glossopharyngeal nerve.

Posterior third of tongue is nodular from underlying lymphoid tissue (*lingual tonsil*) and mucous glands.

Inferior surface of tongue is covered by very much thinner mucous membrane and on each side running forwards the *ranine (sublingual) vein* can be seen.

The *extrinsic muscles* of the tongue are attached to bone—they alter the position and shape of the tongue.

Genioglossus
Origin: Superior tubercle of mental spine (genial tubercle) on inner surface of mandible near symphysis.

Insertion: Body of hyoid bone (posterior fibres); body of tongue from root to tip (anterior fibres). Forms main bulk of the tongue. Separated by midline fibrous septum.

Nerve supply: Hypoglossal.

Action: Inferior fibres raise hyoid bone, draw tongue forwards and protrude it to opposite side. Anterior fibres withdraw tip of protruded tongue.

Hyoglossus

   Origin: From side of body and greater and lesser horns of hyoid.

   Insertion: Back and side of tongue, interdigitating with styloglossus.

   Nerve supply: Hypoglossal.

   Action: Depresses sides of tongue, making surface convex transversely.

   Structures deep to hyoglossus: Inferior longitudinal muscle of tongue, genioglossus and middle constrictor muscles, lingual vessels, stylohyoid ligament below, glossopharyngeal nerve above.

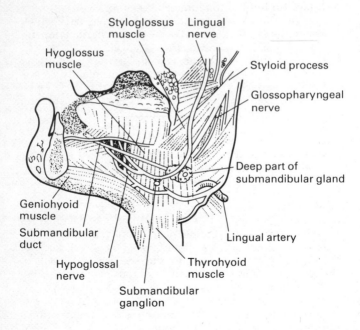

Fig. 69 Floor of mouth left side exposed by removal of left half of mandible and left mylohyoid muscle.

Superficial relations of hyoglossus: All structures in floor of mouth (Fig. 69). These include the deep part of the sub-mandibular gland with its duct passing forwards, the lingual nerve lateral to, below and then medial to the duct, the submandibular ganglion attached to the lingual nerve and the hypoglossal nerve running forwards below the lingual.

Styloglossus

Origin: Lateral surface of apex of styloid process and stylohyoid ligament.

Insertion: Side of dorsum of tongue, interdigitating with hyoglossus.

Nerve supply: Hypoglossal.

Action: Draws tongue upwards and backwards; makes superior surface concave transversely.

Palatoglossus (in palatoglossal arch)

Origin: Inferior and lateral surface of soft palate.

Insertion: Side and dorsum of tongue.

Nerve supply: Pharyngeal plexus (p. 262).

Action: Approximates back of tongue to soft palate. The *oropharyngeal isthmus*, opening between mouth and oropharynx, is bounded by the soft palate, the palatoglossal folds and the back of the tongue.

The *intrinsic muscles* are entirely within the substance of the tongue, alter its shape and are all supplied by hypoglossal nerve.

Superior longitudinal muscle

One on each side. Longitudinal fibres lying under the mucous membrane. It arises from the median glosso-epiglottic fold and from the midline septum; the fibres pass obliquely outwards, the anterior fibres being longitudinal to the side of the tongue.

Inferior longitudinal muscle
Bundle of muscular fibres running along the inferior surface of the tongue from base to tip on each side and changes direction of tip. It lies between the genioglossus and hyoglossus muscles. It arises from the septum at the base of the tongue and passes to the tip.

Transverse muscle
Forms a horizontal layer of muscular fibres between the superior and inferior longitudinal muscles and makes tongue longer and narrower. The fibres are attached to septum and pass laterally to the sides of the tongue.

Vertical muscle
Arises from dorsum, mingles with transverse fibres and makes tongue shorter and wider.

Septum
A vertical fibrous partition extending, in the muscular portion, from the hyoid bone to the tip.

The nerve and blood supply of the tongue
   Motor: Hypoglossal except palatoglossus which is supplied by pharyngeal plexus.
   Sensory: Anterior two-thirds: general sensation—lingual nerve from mandibular; taste—chorda tympani from facial. *Posterior one-third:* general sensation and taste—glossopharyngeal.
   Main *artery* is lingual from external carotid. Also tonsillar branch of facial. *Veins,* superficial and deep to hyoglossus, go to internal jugular.

THE PALATE

The palate forms the roof of the mouth, and consists of the *hard palate* (anterior three-quarters) and the *soft palate* (posterior one-quarter).

The *hard palate* consists of the palatine processes of the maxillae anteriorly, and horizontal plates of the palatine bones posteriorly, together with the mucous membrane and the periosteum adherent to them (*mucoperiosteum*). The greater palatine artery and nerve run forwards (artery lateral to nerve) near the alveolus. There are many small mucous glands in the mucoperiosteum.

The *soft palate*, consisting of muscles, aponeurosis, vessels, nerves, etc., enclosed between two layers of mucous membrane, is attached in front to the posterior margin of the hard palate. The sides blend with the pharynx. From the middle of the posterior edge the *uvula* projects. On each side the soft palate passes into two folds of mucous membrane enclosing muscular fibres—the anterior *palatoglossal* and posterior *palatopharyngeal arches*, between which the *palatine tonsil* lies. The narrowed passage between the anterior arches, leading from the mouth to the pharynx is called the *oropharyngeal isthmus (isthmus of the fauces)*. Arches are also called *anterior* and *posterior pillars of fauces*.

The *aponeurosis of the soft palate*, attached to the crest of the palatine bone and the posterior edge of the bony palate, is formed by the tendon of the tensor veli palatini after it hooks around the hamulus of the medial pterygoid plate. Oral surface of aponeurosis is covered by mucous glands that extend back into uvula.

Muscles of the soft palate

Tensor veli palatini

Origin: Scaphoid fossa at base of medial pterygoid plate, cartilaginous part of auditory tube and spine of sphenoid.

Insertion: Tendon winds round hamulus and passes above buccinator to form broad palatine aponeurosis attached to posterior border of hard palate.

Nerve supply: Mandibular nerve by twig passing through otic ganglion.

Levator veli palatini

Origin: Lower surface of apex of petrous bone, and medial surface of cartilaginous part of auditory tube.

Insertion: Upper surface of palatine aponeurosis.

Nerve supply: Cranial accessory through pharyngeal branch of vagus to pharyngeal plexus.

Azygos uvulae: Two small slips lying on either side of midline in substance of uvula.

The *palatoglossus* and *palatopharyngeus* forming the faucial pillars (arches of oropharyngeal isthmus) are described with the tongue (p. 329) and pharynx (p. 338).

Movements of soft palate

Tensor renders the aponeurosis more tense and thereby flattens and slightly depresses it. *Levator* raises the tensed palate to shut off the oropharynx from the nasopharynx in swallowing and in speech, except for the nasal consonents. Lower layer of *palatopharyngeus* (p. 338) depresses soft palate to close oropharyngeal isthmus. *Palatoglossus* raises tongue, but its upper attachment is too far forwards to have much effect on soft palate itself.

The *palatine tonsils* occupy the recesses between the arches of the oropharyngeal isthmus, the anterior arch being formed by the palatoglossus and the posterior by the palatopharyngeus. Lateral to the tonsil is the superior constrictor and medially the pharyngeal mucous membrane. Their arterial supply is large, from the ascending pharyngeal, ascending and greater palatine (maxillary), tonsillar (facial) and dorsalis linguae (lingual) arteries.

THE SALIVARY GLANDS

The parotid gland

This is the largest and lies below the external acoustic meatus and between mastoid process behind, and ramus of mandible in front. Anteriorly, it overlaps the hinder part of the masseter. It extends downwards on the sternocleido-mastoid muscle. It is enclosed between two layers of cervical fascia (*parotid fascia*).

The deep surface is irregular, and is grooved obliquely by the styloid process and its attached muscles; behind this process a lobe passes medially lying on the internal jugular vein, and glossopharyngeal, vagus and accessory nerves. In front of the styloid process a large lobe (carotid) overlies the internal carotid artery, and lies medial to the ramus and medial pterygoid muscle.

The *accessory parotid gland* is a separate lobe projecting from the anterior border and lying on the masseter.

The *parotid duct* is 5 cm long; comes off from the anterior border crosses the masseter and turns at a right angle to pierce the buccinator and buccal mucous membrane. Its opening in the mouth is on a papilla opposite the 2nd upper molar tooth.

Course of the duct: Middle third of a line from intertragic notch of auricle to middle of upper lip. The duct is palpable. The transverse facial artery lies above the duct and the buccal branch of the facial nerve below.

The parotid gland is traversed by the external carotid artery, retromandibular (posterior facial) vein (the artery is deep to the vein) and, nearer the surface, by the branches of the facial nerve.

Embedded in it superficially are several preauricular lymph nodes.

Nerve supply: Parasympathetic in the glossopharyngeal. The latter branch may be traced as follows: the tympanic branch of glossopharyngeal gives off lesser petrosal which enters and leaves the skull to synapse in otic ganglion; postganglionic branches join the auriculotemporal nerve and supply the gland. *Sympathetic* from plexus around middle meningeal artery (cell bodies in superior cervical ganglion).

The submandibular gland

This lies in submandibular fossa under cover of the body of the mandible in submandibular triangle.

Superficial lobe: Separated by stylomandibular ligament from parotid; grooved above and behind by facial artery. Lies on mylohyoid, stylohyoid, hyoglossus and anterior belly of digastric; covered superficially by platysma, cervical fascia and facial vein; bounded below by tendon of digastric. Has important lymph nodes (submandibular) close to or even embedded in it.

Deep lobe: Passes with duct between mylohyoid and hyoglossus around posterior free border of mylohyoid.

The *submandibular duct* passes with deep lobe of gland on upper surface of mylohyoid on the hyoglossus. At first the lingual nerve lies above the duct and the hypoglossal nerve below. The lingual nerve passes lateral to the duct, then below and then medial to it. The duct passes upwards and forwards on genioglossus to open on the papilla by the side of the lingual frenulum.

Nerve supply: Parasympathetic in chorda tympani synapsing in submandibular ganglion or in gland itself. *Sympathetic* from plexus round facial artery.

The sublingual gland

This lies in the sublingual fossa of mandible under mucous membrane of floor of mouth with its anterior extremity close to the frenulum. The mylohyoid is inferior and the genioglossus medial.

Sublingual ducts (18 to 20): Open separately in the floor of the mouth. One from the posterior part, the larger sublingual duct, opens into or by the side of the submandibular duct.

Nerve supply: Parasympathetic from chorda tympani synapsing in submandibular ganglion.

There are small salivary glands in mucous membrane of lips, cheeks, palate and tongue.

THE PHARYNX

The pharynx is an oval musculomembranous structure attached to base of skull and lying behind the nose, mouth and larynx. Above, where it is a part of both respiratory and alimentary tracts, its walls are rigid and it is always patent; below the level of laryngeal aperture it is normally a transverse slit (*hypopharynx*) which opens on deglutition. The part above soft palate and behind the nasal cavity is the *nasopharynx*, that behind the mouth the oropharynx, below which is the *laryngeal pharynx*. It is about 12 cm long, and extends from the base of the skull to the lower border of the cricoid cartilage in front (same level as lower border of 6th cervical vertebra behind). It is widest at the level of the hyoid bone. Behind, it is separated by the prevertebral fascia from the longus cervicis and capitis muscles of each side. Below, it is continuous with the oesophagus.

Openings of pharynx: The nasopharynx communicates anteriorly with the nasal cavity through the choanae (posterior nasal apertures); the auditory tubes open onto the lateral wall of the nasopharynx just above level of palate; the oropharynx opens into the mouth through the oropharyngeal isthmus; inferiorly there are the laryngeal and oesophageal openings.

The *pharyngobasilar fascia* forms a fascial layer in the wall of the pharynx. It is thin below, but strong above, where it forms part of the wall of the nasopharynx and fills the space above the upper crescentic margin of the superior constrictor over which the auditory tube passes. It is attached above to the basilar part of the occipital bone and petrous temporal, and is strengthened by a process of fascia attached to the pharyngeal tubercle of the occipital bone (the *pharyngeal ligament*). Inferiorly it becomes lost between the muscular and mucous strata.

Recesses of pharynx: In nasopharynx there is a narrow slit between tube and prevertebral muscles, the *pharyngeal recess*. In oral part, between base of tongue and epiglottis, are *valleculae*, separated by the *median glosso-epiglottic fold*.

Each vallecula limited laterally by *lateal glosso-epiglottic fold*.
In laryngeal part of pharynx between laryngeal aperture
and side wall of pharynx on each side there is the *piriform
fossa*.

Muscles of the pharynx

Superior constrictor
   Origin: (1) Posterior border of medial pterygoid plate, (2)
pterygomandibular raphe, (3) posterior end of mylohyoid
line of mandible, (4) side of tongue.
   Insertion: (1) Pharyngeal tubercle at base of skull, (2)
median raphe extending downwards in midline from
pharyngeal tubercle, blending with muscle of opposite side
down to level of vocal folds. The muscle is external to
pharyngobasilar fascia.

Middle constrictor
   Origin: (1) Upper border of greater cornu of hyoid bone,
(2) lesser cornu and lower part of stylohyoid ligament.
   Insertion: Median raphe, overlapping superior constrictor
down to level of vocal folds (Fig. 70).

Inferior constrictor
   Origin: (1) Outer surface of thyroid cartilage behind
oblique line, (2) cricoid cartilage.
   Insertion: Thyroid part into median raphe below level of
vocal folds (muscle wall thin here and may be site of
diverticulum). Cricoid fibres are sphincteric (no raphe) and
named *cricopharyngeus*.

   Nerve supply: The constrictors are supplied by the cranial
accessory nerve (nucleus ambiguus) through the vagus, its
pharyngeal branch and pharyngeal plexus. Cricopharyngeus also supplied by recurrent and external laryngeal
nerves.

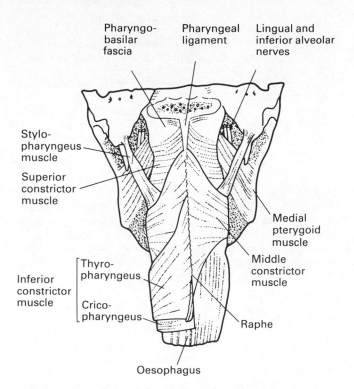

Fig. 70 Posterior view of pharynx (right inferior constrictor has been removed).

Stylopharyngeus

Origin: Root of styloid process.

Insertion: (1) Posterior border of thyroid cartilage, (2) wall of pharynx. (Fig. 70).

Nerve supply: Glossopharyngeal (cell bodies in nucleus ambiguus).

Action: Pulls larynx and pharynx upwards in swallowing.

Palatopharyngeus
Origin: Upper surface of palatine aponeurosis by two layers embracing levator palati. Upper layer from hard palate, lower layer from aponeurosis only.
Insertion: Blends with constrictors; posterior border of thyroid cartilage.
Nerve supply: Cranial accessory (nucleus ambiguus), through pharyngeal plexus.
Action: Pulls larynx and pharynx upwards and closes fauces in swallowing.

Palatopharyngeal sphincter
Formed by soft palate and a U-shaped sling within the superior constrictor with limbs attached on posterior margin of hard palate lateral to upper layer of palatopharyngeus. Contraction of sling raises a ridge (*Passavant's ridge*) which meets the soft palate when it is elevated in swallowing. Greatly hypertrophied in cleft palate.

Salpingopharyngeus
Origin: Medial end of cartilaginous tube.
Insertion: Thyroid cartilage and side wall of pharynx. Produces a ridge (*salpingopharyngeal fold*) in mucosa inferior to tubal elevation.
Action: Pulls larynx and pharynx upwards in swallowing; at same time opens cartilaginous part of auditory tube (latter also due to tensor and levator veli palatini).

THE OESOPHAGUS

The oesophagus extends from pharynx to stomach, and is 25 cm long. It begins at the lower border of cricoid cartilage opposite the 6th cervical vertebra and ends at the level of the 11th thoracic vertebra. The distance from the front teeth to the junction of oesophagus and stomach is 40 cm.

Course and relations
In the neck: Passes downwards and slightly to the left. The

trachea and thyroid gland are anterior and the vertebrae covered by longus colli and prevertebral fascia are posterior. The common carotid artery is lateral and the recurrent laryngeal nerve is between the oesophagus and trachea. On the left the thoracic duct is posterior. The inferior thyroid artery runs with the recurrent laryngeal nerve.

In the thorax: In the superior mediastinum—passes downwards to the right to reach midline at level of 5th thoracic vertebra. *In the posterior mediastinum*—passes forwards, downwards and to the left, with the vagus nerves which form a plexus on its surface; the left nerve is anterior and the right posterior.

Superiorly the trachea and its bifurcation are anterior, and inferiorly the heart (left atrium) and pericardium are anterior. Posteriorly is the vertebral column. As the oesophagus passes to the left near the diaphragm it lies in front of the aorta which is on its left higher up. The vena azygos is on its right and the hemiazygos veins on its left. The thoracic duct is posterior. The arch of the aorta and the left bronchus as they pass to the left are anterolateral. The right posterior intercostal arteries and hemiazygos veins pass to the right behind the oesophagus. The left recurrent laryngeal nerve runs upwards between the oesophagus and trachea.

In the abdomen: Passes through oesophageal opening in right crus of diaphragm at level of the left 6th costal cartilage and 10th thoracic vertebra, to end at the cardiac opening of the stomach. Accompanied by vagus nerves. Lies behind left lobe of liver.

Structure of the oesophagus: Three coats: (1) mucous membrane lined by stratified squamous epithelium, (2) submucous, areolar tissue with many mucous glands, (3) muscular, two layers; inner circular, outer longitudinal; both layers consist of striated muscle above but soon contain only smooth muscle.

Nerve supply: Upper third recurrent laryngeal nerves and sympathetic from middle cervical ganglion via inferior thyroid artery. Below lung roots, vagus nerves and sympathetic in oesophageal plexus.

Arteries: Oesophageal from inferior thyroid, thoracic aorta and right intercostals, and at lower end left gastric.

Note: Left gastric vein (a portal tributary) anastomoses in lower oesophagus with systemic veins.

THE STOMACH

The form and position of the stomach in the living subject is unlike the flaccid bag seen in the cadaver. In the living, when not distended, the stomach is contracted and its position varies with posture and with the degree of distension of the intestines. In the erect posture, when containing a small amount of fluid, its form is J-shaped; the vertical stalk of the J represents the upper two-thirds of the stomach (*cardiac part*) which is surmounted by a convex cap, the *fundus,* defined as that part above the cardiac orifice; on the right, below the fundus, the *cardiac orifice* is the opening between the stomach and the oesophagus; the right margin (*lesser curvature*), which is vertical in its upper two-thirds, turns upwards and to the right and limits the *pyloric part;* the angle between the two parts is known as the *angular notch.* The left margin of the cardiac portion is variably convex (*greater curvature*), and inferiorly may reach or lie below the umbilicus; it then passes upwards and to the right as the lower right margin of the pyloric portion.

The lesser omentum is attached to the lesser curvature and the greater omentum to the greater curvature. The greater omentum to the left is continuous with the gastrosplenic ligament.

The cardiac orifice is fixed by the passage of the oesophagus through the diaphragm and lies at the level of the 10th thoracic vertebra and behind a point on the 7th left costal cartilage 2.5 cm from the sternum. The *pylorus* is palpable as a thickening of the circular muscular coat (*pyloric sphincter*), and is also marked by a pyloric vein; it is movable to the right as the stomach distends, but usually lies within 2.5 cm to the right of the midline, opposite the body of the first lumbar vertebra; this lies on the *transpyloric plane* which is a plane through the midpoint of a line joining the jugular notch with the upper border of the pubis. The pylorus lies in contact with the inferior surface of the quadrate lobe of the liver or the neck of the gall-bladder.

The anterior surface is covered by peritoneum of the greater sac and the posterior surface by peritoneum of the lesser sac except near the cardiac orifice.

Relations: Anterior surface, which also looks upwards, is in contact with, from left to right, diaphragm, abdominal wall (epigastric region), inferior surface of liver.

Posterior surface is separated from diaphragm, aorta, pancreas, spleen, left kidney and suprarenal, transverse mesocolon and colon, by omental bursa (lesser sac of peritoneum).

The stomach lies in the epigastric and umbilical regions of the abdomen; when distended it may encroach upon the left hypochondriac region.

Arteries: Right and left gastric run along lesser curvature; right and left gastro-epiploic along greater curvature; short gastric branches from the splenic to fundus.

Nerves: Right vagus to posterior surface; left vagus to anterior surface; sympathetic from the coeliac plexus to both surfaces.

Structure: Four layers: (1) mucous membrane, very rugose, (2) abundant submucous layer, (3) muscular coat, oblique, circular, and longitudinal from within outwards, (4) serous from peritoneum.

THE SMALL INTESTINE

This extends from the pylorus of the stomach to the ileocaecal opening where it joins the large intestine. It is divided into a fixed C-shaped part, the *duodenum* and a mobile part, the *jejunum* and *ileum*, which are suspended from the posterior abdominal wall by the mesentery.

The duodenum

Length: 25 cm.

Shape: Horse-shoe, with the convexity to the right side; the concavity encloses the head of the pancreas.

Position: Lies in epigastric and umbilical regions. Has no mesentery, and is covered by peritoneum (retroperitoneal). Divided into four parts.

RELATIONS

Superior (1st) part

5 cm long; directed from pylorus upwards, backwards, and to the right, reaching the neck of the gall bladder. The 1st half is almost surrounded by peritoneum, but the 2nd is covered only in front.

Anterior and superior: Liver, gall bladder.

Posterior: Common bile duct, portal vein and gastro-duodenal artery.

Superior: Epiploic foramen (aditus to lesser sac) and lesser omentum.

Inferior: Head of pancreas.

Descending (2nd) part

8 cm long; vertical; passes from level of neck of gall bladder to 3rd lumbar vertebra. Common opening of bile and pancreatic ducts enter at *duodenal papilla* on posteromedial wall. Covered in front by peritoneum, except where crossed by transverse colon.

Anterior: Transverse colon, liver and gall bladder, small intestine.

Posterior: Right kidney, suprarenal gland, renal vessels, and inferior vena cava.

Left side: Head of pancreas, bile and pancreatic ducts.

Right side: Right colic flexure.

On the inner aspect, 10 cm from pylorus, is *duodenal papilla* on which is orifice for both bile and pancreatic ducts.

Inferior (horizontal, 3rd) part
About 10 cm long; passes from right to left across 3rd lumbar vertebra; ends on left side of 3rd and lies inferior to transverse mesocolon; is covered in front by peritoneum, except where root of mesentery crosses it.

Anterior: Superior mesenteric vessels and autonomic plexus.

Posterior: Inferior vena cava, crura of diaphragm, aorta, left psoas and left renal vessels.

Superior: Pancreas.

Ascending (4th) part
About 3 cm long. Ascends vertically on left psoas to side of 2nd lumbar vertebra, turns forward to join jejunum.

Anterior: Peritoneum, jejunum and transverse colon.

Posterior: Left gonadal artery, sympathetic trunk and psoas.

Medial: Pancreas.

Duodenojejunal flexure
This lies to left of 2nd lumbar vertebra where the small intestine forms an acute bend forwards, downwards and to the left; attached to the right crus of the diaphragm by a band of unstriped muscle, known as the *suspensory muscle of duodenum (Treitz)*. To the left of the flexure is seen the *paraduodenal fossa* with the inferior mesenteric vein forming its prominent left margin.

Arteries: Right gastric, superior pancreaticoduodenal from gastroduodenal, inferior pancreaticoduodenal from superior mesenteric.

Nerves: From coeliac plexus (vagal and sympathetic).

The jejunum and ileum

The jejunum forms two-fifths of the rest of the small intestine which is up to 23 feet long when removed from a cadaver; commences on the left side of the 2nd lumbar vertebra and continues into the ileum; it is wider, and its coats are thicker, more vascular and of a deeper colour than those of the ileum.

The ileum consists of the remaining three-fifths of the small intestine and terminates in the right iliac fossa by opening into the caecum; some coils about 30 cm from ileocaecal valve lie in pelvis.

The ileum with the jejunum is suspended from the posterior abdominal wall by the *mesentery* (p. 359). Their vessels are derived from the superior mesenteric artery and the veins drain into the vein of the same name which joins the splenic vein to form the portal vein (p. 198).

Structure of small intestine: (1) Mucous membrane, lined by columnar epithelium and raised, especially in duodenum, into *plicae circulares*; a prominent plica surmounts the duodenal papilla, and from its lower margin a vertical plica runs downwards for a short distance; there are villi and intestinal glands throughout; *aggregated lymphatic follicles (Peyer's patches)*, of oblong shape, with long axis along that of bowel, are found on antimesenteric aspect, chiefly in lower ileum; (2) submucous; (3) muscular, circular and longitudinal; (4) serous, incomplete over duodenum, elsewhere complete except at mesenteric border.

The duodenum is wider and has a thicker coat than the rest of the small intestine. It has well marked plicae circulares which are absent in its superior part. The plicae are slight or absent in the ileum.

THE LARGE INTESTINE

Extends from the ileum to the anus and is about 1.5 m long. The main features distinguishing the large from the small intestine are (a) the longitudinal muscle in the large intestine is in the form of three bands (*taeniae*), unlike the continuous layer in the small, (b) the *sacculations (haustra)* of the large bowel due to the longitudinal muscle being shorter than the rest of the wall, (c) the *appendices epiploicae*, small tags of fat projecting from the wall of the large intestine.

The large intestine consists of the caecum, vermiform appendix, ascending, transverse, descending and sigmoid colons, rectum and anal canal.

Caecum and vermiform appendix
Caecum is dilated blind pouch, inferior to the *ileocaecal opening* situated in the right iliac fossa, and completely covered by peritoneum so that it is mobile with a space behind it in which the appendix is most frequently found (*retrocaecal*—65%). The three taeniae of the caecum converge onto its medial posterior part where the vermiform appendix is attached by a short triangular mesentery (*meso-appendix*) to the mesentery of the terminal ileum. The appendix is a blind tubular projection about 7 cm long and about 8 mm in diameter. Its second commonest position is downwards and medially into the pelvis (pelvic—20%).

The *blood supply* of the appendix is the appendicular artery from the posterior caecal. The caecum is supplied by branches from the ileocolic, a branch of the superior mesenteric.

The *ileocaecal valve* lies on the left side of the junction of the caecum with the colon; is formed by the circular muscle of the ileum passing through the wall of the caecum. The opening is slit-like with prominent upper and lower lips; the lips fuse at either side to form the *frenulum*.

Ascending colon
Extends from the ileocaecal orifice to the inferior surface of the liver to the right of the gall bladder, where it turns to the left at the *right colic (hepatic) flexure*. The peritoneum covers the anterior and lateral surfaces; the taeniae are anterior, medial and posterolateral.

The iliacus, quadratus lumborum and lateral part of right kidney are posterior to the ascending colon.

The *blood supply* of the ascending colon is the right colic artery from the superior mesenteric.

Transverse colon
Passes in the umbilical region from right to left, from the gall bladder to the spleen where it turns downwards at the *left colic (splenic) flexure*. It forms an arch, convex anteriorly and below. It is attached by the *transverse mesocolon* to the

pancreas on the posterior abdominal wall. Above the transverse colon are the liver, gall bladder, greater curvature of stomach and spleen, and below is the small intestine.

Anteriorly there are the anterior layers of the greater omentum and the anterior abdominal wall.

Posteriorly there are the right kidney, descending part of duodenum, pancreas, small intestine and left kidney.

The position of the transverse colon varies so that it may be high enough to lie behind the stomach or low enough to reach the pelvis.

The *blood supply* of the transverse colon is from the middle colic, a branch of the superior mesenteric. The artery anastomoses with the upper left colic from the inferior mesenteric at the left colic flexure.

Descending colon

Passes from the left colic (splenic) flexure to the sigmoid (pelvic) colon; the calibre of the bowel here is very small. At the level of the 10th left rib a fold of peritoneum, the *phrenicocolic ligament*, passes from the left flexure to the diaphragm. The colon then passes downwards to the pelvic brim and ends in the sigmoid colon. The peritoneum of the descending colon invests its anterior and lateral surfaces.

Posteriorly are the left kidney, quadratus lumborum and iliacus, anteriorly the small intestine and medially, lateral border of left kidney.

The *blood supply* of the descending colon is from the left colic artery from the inferior mesenteric.

Sigmoid (pelvic) colon

Extends from the medial border of the psoas to the level of the 3rd sacral vertebra. It has an extensive mesentery (*pelvic mesocolon*) and forms a loop hanging into the pelvis. As its lower end is reached the longitudinal muscular coat becomes uniformly spread out over its anterior and posterior surfaces.

The pelvic mesocolon is attached by two limbs—one to the brim of the pelvis lateral to the left sacro-iliac joint and the other to the upper part of the sacrum from the joint to the middle of the 3rd sacral vertebra. The apex of the two limbs is in front of the left ureter as it lies anterior to the division of the common iliac artery.

The sigmoid colon hangs into the pelvis. Its lateral relations are the left ovary (or ductus deferens) and obturator nerve, and posteriorly are the internal iliac vessels, sacral plexus and piriformis. The pelvic colon lies on the bladder and uterus.

The *blood supply* of the sigmoid colon is from the sigmoid arteries from the inferior mesenteric.

Rectum
Commences at 3rd sacral vertebra as continuation of pelvic colon; passes downwards in concavity of sacrum; at coccyx turns forwards as dilated *ampulla*; about 3 cm in front of the coccyx it bends sharply backwards to become the anal canal.

It has peritoneum on the upper third of its anterior and lateral surfaces and on the middle third of its anterior surface. The lower third is inferior to the peritoneum.

Relations: Anterior, in male, rectovesical pouch; triangular area at base of bladder, seminal vesicles, ductus deferentes, posterior surface of prostate; *in female,* posterior wall of vagina and recto-uterine pouch.

Posterior, sacrum, piriformis muscles, sacral plexuses, coccyx, levatores ani.

Lateral, pararectal fossa, inferior to coccygeus muscle.

The rectum has three lateral bends. The highest and the lowest are concave to the left; they are produced by the relative shortness of the longitudinal muscular coat which covers the whole circumference of bowel. Internally they form horizontal shelves of mucosa (*horizontal folds of rectum, valves of Houston*). In its lowest part the rectal cavity is dilated as the *rectal ampulla*.

Structure of large intestine
(1) Mucous membrane smooth, with columnar epithelium and mucous crypts; (2) Submucosa; (3) Muscular, internal circular, outer longitudinal, latter arranged in three bands (taeniae) along colon but uniform in rectum; (4) Serous, variable (for individual peculiarities see different parts).

Anal canal

Extends from pelvic floor to anus and is about 3 cm long; directed downwards and backwards.

Relations: Anterior, in male, membranous part of urethra, bulb of corpus spongiosum; *in female*, posterior wall of vagina, with perineal body (central tendon of perineum) intervening.

Lateral and posterior: Levatores ani, which, uniting, support it as in a sling; ischiorectal fossa.

Anorectal sphincters (p. 127).

Upper half or more of anal canal is true gut, lined by columnar epithelium; shows *rectal columns (of Morgagni)* produced by blood vessels. Lower half or less of canal is ectodermal in origin and lined by skin. At junction of two regions are *anal valves* connecting lower ends of columns, and *white line* representing junction of skin with mucous membrane.

THE LIVER

This is the largest gland in the body and is situated in the upper right part of abdominal cavity immediately below diaphragm in right hypochondriac, epigastric and part of left hypochondriac regions. Its average weight is 1.5 kg.

Relations

Superior surface is convex and covered by peritoneum; above is the diaphragm; divided descriptively into two unequal lobes (right and left) by a sagittal fold of peritoneum, the *falciform ligament*.

Inferior surface (Fig. 71) is concave and slopes facing postero-inferiorly; is in contact with the stomach (left lobe), pylorus, duodenum, right colic flexure, right kidney and right suprarenal gland (right lobe); is covered with peritoneum, except where gall bladder is attached, and where lesser omentum is attached at the porta hepatis and the fissure for ligamentum venosum.

The upper part of the inferior surface is broad and round and is connected to the diaphragm over the right lobe by the coronary ligament between the two layers of which its surface is non-peritoneal (*bare area*). To left of bare area at level of 10th and 11th thoracic vertebrae, is the caudate lobe, which bounds the lesser sac in front. Right suprarenal, inferior vena cava, aorta, oesophagus and lesser peritoneal sac lie behind.

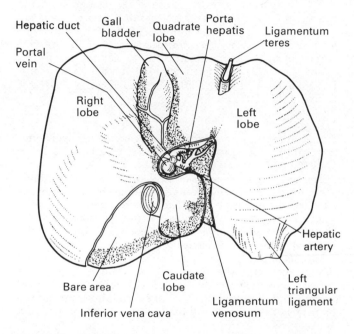

Fig. 71 Inferior surface of liver.

Anterior surface is triangular and marked by a notch on its lower border where the falciform ligament is attached; in contact with diaphragm and anterior abdominal wall.
Right surface is convex and in contact with diaphragm.

Ligaments
Five in number; four are composed of peritoneum.

Falciform ligament, double fold of peritoneum, sickle-shaped, in sagittal plane, attached to diaphragm and anterior abdominal wall as far as umbilicus; behind to anterior surface of liver; inferior free edge encloses the ligamentum feres.

The *triangular ligaments*, *right* and *left*, extend from the sides of the diaphragm to the posterior border of the liver.

The *coronary ligament* is continuous with the right triangular ligament and attaches the upper part of the inferior surface of right lobe of the liver to the diaphragm.

The *ligamentum teres* is obliterated left umbilical vein and is contained within posterior or free edge of falciform ligament (Fig. 71).

Fissures
Longitudinal fissure divides the organ descriptively into right and left lobes; is separated into two parts by its union with the porta hepatis.

The anterior part or *fissure for ligamentum teres* contains the remains of left umbilical vein and lies between the left lobe and the quadrate lobe. The posterior part or *fissure for ligamentum venosum* lies between the left lobe and the caudate lobe and contains the ligamentum venosum, the remains of the ductus venosus.

Porta hepatis is placed at right angles to the longitudinal fissure, between the quadrate and caudate lobes, and contains the hepatic ducts, hepatic artery, portal vein, nerves and lymphatics. The ducts are in front of the branches of the artery and the branches of the portal vein are posterior.

Fissure for the inferior vena cava is placed obliquely at the posterior margin of the liver, behind and superior to the gall bladder; lies between the right lobe and the caudate lobe and is separated from the porta hepatis by the *caudate process* (Fig. 71). The hepatic veins enter the vena cava in the depths of this fissure. They have no extrahepatic course.

Lobes

Right and left lobes are separated from each other by fissure for ligamentum teres on inferior surface and posteriorly by the fissure for the ligamentum venosum. The right is the larger, and contains the porta hepatis and fissure for the inferior vena cava; is subdivided into three.

The *quadrate lobe* is bounded by the fissure for ligamentum teres and porta hepatis, and the fossa for the gall bladder.

The *caudate lobe* projects between the fissure for the inferior vena cava and ligamentum venosum behind the porta hepatis.

The *caudate process* connects the preceding lobe with the main mass of the right lobe and lies behind the porta hepatis.

Note: Above division into right and left lobes is descriptive only; actually vascular input and bile output are in almost two equal halves; quadrate lobe and most of caudate lobe belong functionally to left half of liver.

The *fossa for the gall bladder* lies on the inferior surface of the right lobe, parallel to the fissure for ligamentum teres, and separates the quadrate lobe from the main mass of the right lobe.

Vessels of the liver

The hepatic artery from the coeliac artery enters the porta hepatis and divides into two branches for the right and left lobes. The right branch runs behind the cystic duct and gives off the *cystic artery* to the gall bladder. Its subsequent branches accompany the branches of the portal vein.

The *portal vein*, behind the hepatic artery (left) and bile duct (right) ascends between the layers of the lesser omentum, in front of the opening of the lesser sac, to the porta hepatis. The portal vein and hepatic artery break up into *interlobular vessels* from which blood passes into the sinusoids between the columns of liver cells within the lobules. The blood is then collected in the centre of the lobules by the *intralobular veins*, which are tributaries of the hepatic veins collecting blood from the liver.

The *hepatic veins* pass out of the liver in the depths of the fissure for the inferior vena cava and immediately join that vessel. There are three, middle, right and left.

THE GALL BLADDER AND BILE DUCTS

The gall bladder is a conical bag placed in a fossa on the inferior surface of the right lobe of the liver. Its long axis is directed upwards, backwards and to the left, from the *fundus* to the *body* and *neck*. Its anterior (superior) surface is attached to the liver, and its fundus and posterior (inferior) surfaces are invested by peritoneum reflected from the adjacent surface of the liver.

Relations: The body is in contact in front with the liver and behind with the superior (1st) part of duodenum and the right flexure of colon. The fundus is in contact with the anterior abdominal wall at the tip of the 9th right costal cartilage where the linea semilunaris (lateral edge of rectus abdominis muscle) meets the costal margin.

The *cystic duct* passes from the neck of the gall bladder downwards, backwards and to the left and joins the hepatic duct at or below the porta hepatis.

The *common hepatic duct*, formed by union of ducts from right and left lobes, issues from the liver at the porta hepatis. It is joined by the cystic duct and forms the *bile duct*. The common hepatic duct is about 2–3 cm long and lies entirely within the portal fissure.

The *(common) bile duct* is formed by the union of the common hepatic and cystic ducts. It passes downwards in front of the opening of the lesser sac between the layers of the lesser omentum, with the portal vein behind and the hepatic artery on the left. It then descends behind the superior part of the duodenum, and, passing between the pancreas and descending part of the duodenum, where it lies on the inferior vena cava, it enters the descending part of the duodenum obliquely on its posteromedial wall by an opening on the duodenal papilla (p. 342) common to it and the pancreatic duct. Length is about 8 cm, one-third in lesser omentum, second third behind superior part of duodenum, last third behind or in head of pancreas.

THE PANCREAS

This organ consists of an enlarged *head* to the right in the
concavity of the duodenum, a narrow part, the *neck*, a *body*
which extends to the left ending in a *tail* which reaches
hilum of spleen. It is situated in epigastric and left
hypochondriac regions and is directed nearly horizontally
across posterior wall of abdomen.

Relations

Head: Embraced by three parts of duodenum, from which
it is partly separated, behind by bile duct and in front by
pancreaticoduodenal arteries. Is covered in front by transverse colon. The *uncinate process* of the pancreas projects to
the left behind the superior mesenteric vessels. Behind the
head are inferior vena cava and right renal vessels.

Neck: Extends from front of head forwards and to left to
merge into body; behind neck is junction of the superior
mesenteric with the splenic vein to form the portal vein.

Body: Superior surface: stomach and lesser omentum;
covered by peritoneum. Posterior surface: aorta, crura of
diaphragm, superior mesenteric vessels, splenic vein, left
kidney, left suprarenal gland and left renal vessels. Superior border: from right to left—coeliac artery, splenic artery.
Inferior surface: from right to left—duodenojejunal flexure,
left colic flexure and small intestines; covered by peritoneum.

Ducts: Main duct extends transversely from left to right,
opens into descending (2nd) part of the duodenum in
common with bile duct. Common part is dilated (*ampulla of
Vater*) and surmounted by a *sphincter (of Oddi)* which
extends upwards round terminal part of bile duct.

Accessory duct drains from lower part of head upwards
and to right, and opens independently into duodenum
above duodenal papilla. It may or may not communicate
with the principal duct.

Arteries: Pancreatic from splenic, superior pancreaticoduodenal from hepatic and inferior pancreaticoduodenal
from superior mesenteric.

Veins: Open into splenic and superior mesenteric.

Nerves: From the coeliac plexus.

THE PERITONEUM (Figs. 72 and 73)

The peritoneum is the serous membrane of the abdominal cavity. It consists of a *parietal layer* lining the abdominal and pelvic walls, and a *visceral layer* reflected more or less over the contained organs. In the male, it is a closed sac, but in the female the free extremities of the uterine tubes open into the cavity. Its disposition can best be learned by tracing its continuity in the uninjected cadaver.

When followed horizontally from the midline below the umbilicus, the peritoneum lines the right half of the abdominal wall, as far as the lumbar region, where it entirely surrounds the caecum and vermiform appendix. It invests only the front and sides of the ascending colon, though occasionally the whole circumference of the gut is enclosed, a mesocolon being then formed. The peritoneum then passes medially, covering the lower part of the anterior

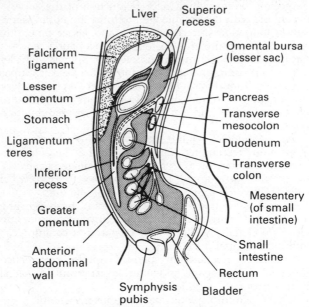

Fig. 72 Sagittal section of peritoneal cavity.

surface of the right kidney, the front of the inferior (3rd) part of duodenum, and goes downwards over the front of the vessels of the small intestines, encloses the small intestines, and is reflected upwards on the posterior surface of the vessels to the vertebrae, thus forming the *mesentery*. From the spine it may be traced to the left over the lower part of the anterior surface of the left kidney to the descending and sigmoid colon, and thence on to the abdominal wall to the midline. The peritoneum of the descending colon is similar to that of the ascending.

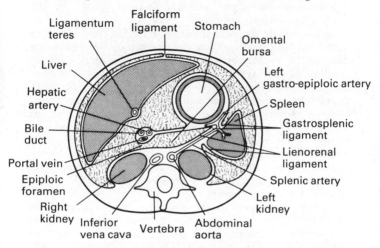

Fig. 73   Horizontal section of peritoneal cavity.

The peritoneum of the pelvis is continuous with that of the abdominal cavity. It encloses completely the sigmoid colon, and forms the *sigmoid mesocolon*. The peritoneum is applied to the front and sides, and lower down to the front only of the rectum, whence it is reflected in the male on to the base and upper part of the bladder, forming the *rectovesical pouch*; this is bounded on each side by a fold of peritoneum passing forwards from the sacrum lateral to the rectum to the bladder to form the *posterior false ligament of the bladder*.

From the apex of the bladder the peritoneum passes upwards on to the anterior abdominal wall, enclosing the remains of the urachus and constituting the *median umbilical fold of the bladder*; laterally it forms a fold along the line of the obliterated umbilical artery to the sides of the pelvis, forming the *medial umbilical fold*. Where the obliterated umbilical artery passes between the abdominal wall and the side of the bladder it raises the peritoneum into a fold, which separates two shallow fossae, the *medial* and *lateral inguinal fossae*. The lateral fossa is bounded by a fold of peritoneum, the *lateral umbilical ligament*, raised by the underlying inferior epigastric artery.

In the female the peritoneum is reflected from the sides and front of the rectum on to the upper part of the posterior wall of the vagina, and thence over the posterior, upper and anterior surfaces of the uterus to the bladder. Between the uterus and rectum is the *recto-uterine pouch (of Douglas)*, and between the uterus and bladder there is the *uterovesical pouch*. The peritoneum passes off from the lateral margins of the uterus to the pelvic wall, forming the *broad ligaments*, in the upper border of which the uterine tubes run; the peritoneum is continuous with their open fimbriated extremities. The free upper margin of the broad ligament lateral to the uterine tubes forms the *suspensory ligament of the ovary (infundibulopelvic ligament)*.

In the upper part of the abdomen the peritoneum is attached to the inferior surface of the diaphragm as far backwards as the posterior surface of the liver and the cardiac orifice of the stomach. It is then reflected forwards on the upper surface of the liver, forming the ligaments of that organ, and passing round the anterior border it is applied to the inferior surface as far as the porta hepatis, where, meeting a peritoneal layer from the posterior surface (from the lesser sac), the two descend to the stomach to form the *lesser omentum*.

The peritoneum passes to the right from the longitudinal fissure of the liver and invests the gall bladder, the inferior surface of the right lobe, and the front of the descending part of the duodenum. It passes to the anterior surface of the right kidney, where it becomes continuous with the part already traced. To the left of the longitudinal fissure, the peritoneum covers the left lobe of the liver, and is reflected over the front of the cardiac end and fundus of the stomach, whence it passes off to invest the spleen, forming one layer of the *gastrosplenic ligament*. From the spleen it is continued over the anterior surface of the left kidney, forming the posterior layer of the *lienorenal ligament*, and over the descending colon to join the part already described.

Passing to the left behind the stomach there is a diverticulum of the main peritoneal cavity (greater sac). The diverticulum is called the *omental bursa (lesser sac)* and extends upwards behind the liver (region of caudate lobe) and downwards into the greater omentum. The opening into the bursa is called the *epiploic foramen*.

Anterior to the opening is the *lesser omentum* containing the portal vein (posterior) the bile duct (anterior and right) and hepatic artery (anterior and left). Posterior to the opening are the inferior vena cava and right crus of diaphragm. It is bounded above by the caudate process of the liver and below by the superior part of the duodenum.

The two sacs traced vertically

From porta hepatis two layers pass to lesser curvature of stomach where they separate, one passing in front and the other behind stomach, thus enclosing it.

These layers join together again at greater curvature, forming anterior layers of greater omentum. They then pass down in front of and beyond transverse colon, bend upwards and backwards, and separate to enclose transverse colon. They continue together to the posterior abdominal wall to the lower border of the pancreas, where they part, one layer passing upwards and the other downwards.

The ascending layer passes over upper surface of pancreas and posterior part of diaphragm and then on to posterior surface of liver to the porta hepatis.

The descending layer passes along superior mesenteric vessels, round jejunum and ileum, and back to posterior abdominal wall, forming the mesentery. It then passes downwards in front of the posterior abdominal wall over the lower part of the aorta and sacral promontory. It surrounds the sigmoid colon and forms sigmoid mesocolon.

The peritoneum then passes forwards in the male to the bladder, forming *rectovesical pouch* (p. 355), and in the female to the vagina and uterus, forming *recto-uterine pouch*. It then passes over uterus to bladder, forming *uterovesical pouch*.

The peritoneum passes over bladder to anterior abdominal wall on to inferior surface of diaphragm, is reflected over upper surface of liver and passes round anterior border of liver to its inferior surface as far as porta hepatis.

The two sacs traced horizontally, at the level of the opening into lesser sac

From falciform ligament of liver, left layer, the peritoneum passes forwards on to abdominal wall and diaphragm, then to left and posteriorly over lateral part of left kidney. It is then reflected forwards behind splenic vessels to spleen, forming posterior layer of lienorenal ligament and then passes round spleen as far as hilum and then to stomach, forming anterior layer of gastrosplenic ligament.

It continues over anterior wall of stomach into anterior layer of lesser omentum and turns round bile duct, forming anterior edge of epiploic foramen and then the anterior wall of omental bursa.

Passing from right to left the peritoneum forms posterior layer of lesser omentum, passes over posterior surface of stomach and forms posterior layer of gastrosplenic ligament, reaching hilum of spleen. It then forms anterior layer of lienorenal ligament.

Passing from left to right, the peritoneum continues over left kidney, aorta and inferior vena cava, and forms posterior boundary of epiploic foramen where bursa ends at opening into peritoneal cavity (greater sac). It then passes over right kidney to inferior surface of diaphragm and deep surface of abdominal wall to right layer of falciform ligament. It can be followed backwards on to liver and then over inferior surface of liver to right border round which it may be traced over anterior surface to falciform ligament.

Processes of the peritoneum

Omenta

Layers of peritoneum connected to the stomach.

Lesser omentum: Double layer from porta hepatis and fissure for ligamentum venosum on liver to lesser curvature of stomach and superior (1st) part of duodenum. Right free border forms anterior boundary of epiploic foramen (aditus to lesser sac). Left border attached to diaphragm between its oesophageal and caval openings.

Greater omentum: From greater curvature of stomach downwards to lower free edge where turns upwards and then behind transverse colon to posterior abdominal wall along middle of pancreas. Contains inferior recess of omental bursa which passes in front of transverse colon.

Mesenteries

Double folds connecting intestine to posterior abdominal wall and containing between the layers blood vessels, lymphatics and nerves of the intestine.

The mesentery: Attached to posterior abdominal wall; beginning on left side of 2nd lumbar vertebra (duodenojejunal flexure) passing downwards across vertebrae to right sacro-iliac joint. Crosses in front of horizontal part of duodenum, aorta, inferior vena cava and right ureter. Contains lymphatics, superior mesenteric vessels, autonomic nerves, jejunum and ileum between its layers. Upper layer is continuous with inferior layer of transverse mesocolon, lower layer with peritoneum on posterior abdominal wall.

Transverse mesocolon: Formed by division of posterior layer of greater omentum round the transverse colon; at the attachment posteriorly on the anterior surface of the pancreas the two layers separate, as explained, into ascending and descending layers (p. 357).

Sigmoid mesocolon: Double layer attached to brim of pelvis to left of left sacroiliac joint and to upper half of hollow of sacrum.

Ligaments

The peritoneum, as explained, is reflected from the abdominal walls to viscera, forming so-called ligaments.

Ligaments of the liver: Falciform—sickle-shaped fold in sagittal plane passing from anterior abdominal wall to upper and anterior surfaces of liver. The inferior free border contains obliterated left umbilical vein (ligamentum teres of liver). *Coronary*—connects right lobe to diaphragm; consists of two layers; the anterior is part of greater sac and the posterior is continuous with peritoneum of superior recess of lesser sac. *Triangular*—the *right* is the right end of the coronary ligament. The *left* is formed by two layers of the greater sac.

Ligaments (false) of the bladder: Posterior ligaments—the edges of the rectovesical pouch (*sacrogenital folds*). *Lateral umbilical folds*—ridges raised by inferior epigastric vessels on anterior abdominal wall. *Medial umbilical folds*—ridges from sides of bladder raised by obliterated umbilical arteries. *Median umbilical fold*—ridge in midline from bladder to anterior abdominal wall as far as umbilicus due to urachus.

Ligaments of the uterus: Broad ligament—one on each side passing from side of uterus to lateral pelvic wall; contains between its folds the round ligament of uterus, uterine tube, the ovary and its ligament, and branches of ovarian and uterine vessels. *Recto-uterine folds*—margins of recto-uterine pouch. *Suspensory ligament of ovary (infundibulopelvic ligament)*—lateral part of the upper margin of broad ligament. Extends from infundibulum of uterine tube and tubal end of ovary to lateral pelvic wall; in its margin are the ovarian vessels, nerves and lymphatics.

Other ligaments: Lienorenal—passes from left kidney to spleen; the anterior layer is formed by the lesser sac, and the posterior by the greater sac. Contains splenic vessels and tail of pancreas. *Gastrosplenic*—passes from stomach to hilum of spleen; anterior layer is greater sac and posterior layer is lesser sac; contains the left gastro-epiploic and short gastric vessels. Below it is continuous with the greater omentum. *Phrenicocolic*—passes from left colic flexure to diaphragm near 11th and 12th ribs.

10
The Respiratory System

THE NOSE AND PARANASAL AIR SINUSES

The nose consists of external and internal parts. The sinuses lie around the nasal cavity and open into it.

THE EXTERNAL NOSE

Partly bony, partly cartilage; nasal bones meet in midline above; *piriform opening* below nasal bones between two maxillae which meet inferiorly to form its lower boundary. *Nasal cartilages* are attached to edge of opening. *Fibrofatty tissue* posteriorly and *major alar cartilage* anteriorly form lateral boundary of *external nasal aperture (naris)*. Alar cartilages turn inwards in midline to form medial boundary of aperture below nasal septum. Skin of external nose extends into nasal cavity for a short distance and is hairy (*vestibule of nose*).

NASAL CAVITY (Fig. 74)

The nasal cavity is divided into two by the septum and opens in front by the nares and behind by the *posterior nasal apertures (choanae)*. Each half of the cavity possesses a roof, floor and medial and lateral walls.

The *roof*, which is arched, is formed anteriorly by the nasal bones and nasal spine of frontal, in the middle by the cribriform plate of ethmoid, and posteriorly by inferior surface of body of sphenoid.

The *floor* consists anteriorly of palatine processes of maxillae (three-quarters) and posteriorly of horizontal plates of palatine bones.

The *medial wall (nasal septum)* consists *anteriorly* of cartilage of septum, crest of nasal bones, nasal spine of frontal, *posterosuperiorly* of perpendicular plate of ethmoid, *posteroinferiorly* of vomer and rostrum of sphenoid, and *inferiorly* of crests of maxillae and palatine bones.

361

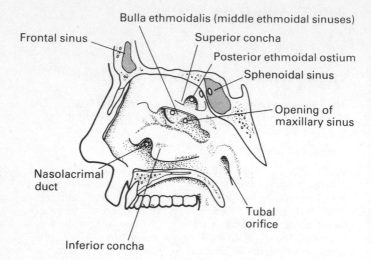

Fig. 74 Lateral wall of nasal cavity, right side (conchae partly removed).

The *lateral wall* is formed mainly by medial wall of maxilla; from in front backwards parts of the lacrimal bone, ethmoid, perpendicular plate of palatine bone and medial pterygoid plate of sphenoid articulate with the maxilla. Inferiorly the inferior nasal concha forms part of the wall.

The *posterior nasal apertures (choanae)* open into the nasopharynx. They are oval in shape and about 2.5 cm vertically and 1 cm transversely.

Projecting medially and then downwards into the cavity from the lateral wall are the *superior* and *middle nasal conchae* of ethmoid and *inferior nasal concha* (a separate bone). Divide lateral wall into four spaces: *spheno-ethmoidal recess* above superior concha and *superior*, *middle* and *inferior meatuses* below the corresponding nasal conchae.

Openings on to lateral wall of cavity
  Spheno-ethmoidal recess: Sphenoidal sinus.
  Superior meatus: Posterior ethmoidal sinuses.

Middle meatus: Contains *bulla of ethmoid* (formed by middle ethmoidal sinuses which open on it) and *hiatus semilunaris* leading into *infundibulum* which receives the *frontonasal duct* from the frontal sinus, anterior ethmoidal sinuses and opening of maxillary sinus.

Inferior meatus: Nasolacrimal duct.

PARANASAL AIR SINUSES

Through the above openings mucous membrane of nasal cavity is continuous with that of paranasal sinuses in the neighbouring bones. Each sinus is lined with ciliated mucous columnar epithelium; direction of ciliary currents is spirally towards opening of sinus.

Sphenoidal sinus
Formed before birth by sphenoidal conchae. Enlarges by excavation of body of sphenoid. Has hypophysis cerebri (pituitary gland) above it and cavernous sinus laterally. Late in life may extend into basi-occiput. Opening in anterior wall into upper back part of nasal cavity.

Ethmoidal sinuses
Anterior, middle and *posterior;* roofed in by the frontal bone. Lie in medial wall of orbit. Anterior may open into infundibulum of frontal sinus or directly into hiatus semilunaris; others open as above.

Frontal sinus
Absent until end of first year. One on each side. When fully developed lies in vertical and orbital plates of frontal bone, forming anteromedial part of roof of orbit. Near frontal lobe of brain. Usually asymmetrical. Open into middle meatus.

Maxillary sinus
Its roof is floor of orbit with infra-orbital nerve. Its floor is alveolar process and extends to lower level than floor of nasal cavity. Roots of upper teeth from 1st premolar to last molar may project into floor. Always 1st molar, usually 2nd premolar or 2nd molar. Posterior and lateral walls have

superior alveolar nerves. Medial wall main part of lateral wall of nasal cavity, with opening large in disarticulated bone made much smaller by encroachment of neighbouring bones (p. 362) and well above floor.

LINING OF NASAL CAVITY

Mucous membrane
Olfactory part is in roof and adjacent parts of medial and lateral walls (septum and superior concha), remainder is respiratory.

The epithelium covering the surface is of three types: (1) stratified squamous in the vestibule of nose, (2) columnar in the olfactory area, (3) ciliated mucous columnar in the respiratory area and paranasal sinuses.

Sensory nerves
Also carry postganglionic parasympathetic secretomotor fibres to mucous glands, mostly from pterygopalatine (sphenopalatine) ganglion (preganglionic fibres are in nervus intermedius of facial nerve). The *septum* is supplied postero-inferiorly by maxillary nerve through nasopalatine (long sphenopalatine) nerves which groove vomer and pass through incisive canal of maxilla to front of hard palate and anterosuperiorly by medial branch of anterior ethmoidal from nasociliary of ophthalmic nerve.

Lateral wall is supplied in quadrants; upper anterior by lateral branch of anterior ethmoidal (ophthalmic), lower anterior by anterior superior alveolar (maxillary), upper posterior by posterior superior nasal (short sphenopalatine) and lower posterior by nasal branches of greater palatine (both maxillary).

Roof of nasopharynx is supplied by pharyngeal branch of maxillary nerve passing through pterygopalatine ganglion, and passing backwards in palatovaginal canal.

The *lining of the sinuses* is supplied by branches of ophthalmic nerve except that of maxillary which is supplied by maxillary nerve.

Olfactory nerves come from cell bodies in olfactory mucous membrane; bundles of olfactory nerves pass through the cribriform plate of ethmoid to reach the olfactory bulb.

Arteries

Facial artery to antero-inferior part of septum; greater palatine from maxillary artery through incisive fossa to septum; otherwise vessels accompany sensory nerves already described (ethmoidal from ophthalmic artery, sphenopalatine from maxillary to lateral and medial walls, infra-orbital and superior alveolar from maxillary). These arteries also supply walls of sinuses.

THE PHARYNX (p. 335)

Although described with the alimentary system, the pharynx obviously is part of the respiratory passages. The nasal cavity leads into the nasopharynx which in turn passes downwards into the oropharynx and then into the laryngeal pharynx which is related to the opening of the larynx. Respiration can also take place through the mouth which leads into the oropharynx.

THE LARYNX

The larynx lies in the front and upper part of the neck below the tongue and hyoid bone and between the large vessels of the neck. It opens above into the pharynx and below into the trachea. In the midline it is related to only skin and cervical fascia but laterally it is covered by the sternohyoid, sternothyroid, thyrohyoid and inferior constrictor muscles. It is composed of cartilages held together by ligaments and intrinsic muscles, and is lined by mucous membrane.

The cartilages of the larynx

Thyroid cartilage

Largest, consisting of two laminae united in front at an acute angle forming the projection known as the *laryngeal prominence (Adam's apple)*; more marked in men than in women and children.

Outer surface: Has oblique line posteriorly passing upwards and backwards and giving attachment to sternothyroid and thyrohyoid and below oblique line to inferior constrictor muscles.

Inner surface: In the midline at the junction of right and left laminae are attached lower end of the epiglottis, the vestibular and vocal folds and the thyro-arytenoid muscles.

Superior border: Sinuous; connected to hyoid bone by thyrohyoid membrane; forms superior thyroid notch above laryngeal prominence.

Inferior border: Forms inferior notch; connected to upper border of cricoid by cricothyroid ligament, and by cricothyroid muscle.

Posterior border: Thick; receives attachment of stylo-, salpingo- and palatopharyngeus muscles; prolonged upwards into *superior horn*, which passes upwards, backwards and medially with lateral thyrohyoid ligament attached to tip, and downwards into *inferior horn*, shorter, passing downwards, forwards and medially; tip has on medial surface a small facet which articulates with the cricoid cartilage by a synovial joint.

Cricoid cartilage
Shaped like signet ring, deep behind(*lamina*) and shallow in front (*arch*).

Outer surface: Anterior half gives attachment to cricothyroid muscles, and behind this to cricopharyngeus part of inferior constrictor. Posterior half, broad and thick, has vertical ridge in midline for attachment of some longitudinal fibres of oesophagus; on each side of ridge a depression gives attachment to posterior crico-arytenoid. At junction of arch and lamina a small facet for articulation with inferior horn of thyroid cartilage.

Inner surface: Smooth, lined with mucous membrane.

Superior border: Inclines from the front upwards and backwards, gives attachment anteriorly to the cricothyroid ligament, and laterally to cricovocal membrane and lateral crico-arytenoid muscle. Upper border of lamina has on each side an articular facet for arytenoid cartilage. Facet slopes downwards and laterally; downward pull of vertical fibres of posterior crico-arytenoid muscle causes descent and therefore separation of arytenoids along these sloping facets.

Inferior border: Horizontal; connected with 1st ring of trachea by cricotracheal ligament.

Arytenoid cartilages

Two small pyramidal cartilages, articulating with upper border of lamina of cricoid; four triangular surfaces.

Anterolateral: Concave and rough; receives attachment of thyro-arytenoid muscle and vestibular fold (false vocal cord).

Posterior: Hollowed for attachment of arytenoid muscle.

Medial: Smooth; covered with mucous membrane.

Base: Has concave facet for articulation with cricoid cartilage forming a gliding synovial joint.

Muscular process: Inferolateral angle; attachment for lateral and posterior crico-arytenoid muscles.

Vocal process: The anteromedial angle, long and pointed for attachment of vocal fold (true vocal cord).

Apex: Directed backwards and medially, surmounted by the *corniculate cartilage* which gives attachment to the aryepiglottic fold.

Cuneiform cartilage

Small cartilage, one on each side in the aryepiglottic fold.

Epiglottis

Projects upwards behind tongue anterior to the superior aperture of the larynx; leaf-shaped lamina of yellow elastic cartilage; behind hyoid bone.

Apex: Attached to angle of thyroid by thyro-epiglottic ligament.

Base: Rounded and free: directed upwards.

Anterior surface: Covered in upper part by mucous membrane which passes forwards in midline as the *median glosso-epiglottic fold* to the tongue; laterally the lateral *glosso-epiglottic (pharyngo-epiglottic) folds* pass to the pharyngeal wall; attached to the hyoid bone by hyo-epiglottic ligament.

Posterior surface: Covered by mucous membrane, concave from side to side, concavoconvex from above downwards; greatest convexity part known as *tubercle* or *cushion of epiglottis.*

The ligaments and membranes

Thyrohyoid membrane
Thickened in the midline as the *median thyrohyoid ligament,* passes from the upper border of the thyroid cartilage to the upper border of the posterior surface of the hyoid bone; between it and the posterior surface of the body of the hyoid bone is a synovial bursa. This membrane forms the anterior wall of the piriform fossa and is pierced by the superior laryngeal vessels and internal laryngeal nerve on each side. *Lateral thyrohyoid ligament* is the posterior edge of the membrane from the superior horn of the thyroid to the tip of the greater horn of the hyoid bone. It contains a small cartilaginous nodule, the *cartilago triticea.*

Cricovocal membrane (conus elasticus)
Attached below to the upper border of the cricoid arch. Passes upwards deep to lamina of thyroid cartilage; its upper free border is attached anteriorly to the deep surface of the laryngeal prominence, and posteriorly, to the vocal process of the arytenoid. Between these attachments it constitutes the *vocal ligament.*

Quadrangular membrane
Encloses the vestibule and separates it from piriform fossa of pharynx. Attached anteriorly to margin of lower half of epiglottis and posteriorly to corniculate cartilage and vocal process of arytenoid. Short free lower margin forms *vestibular fold (false vocal cord).* Long upper margin is also free and forms *aryepiglottic fold.*

The interior of the larynx

The *inlet of the larynx* is the oval, nearly vertical communication with the pharynx. It is bounded in front by the epiglottis, behind by the apices of the arytenoid cartilages on which are the corniculate cartilages, and laterally by the aryepiglottic folds.

The cavity of the larynx extends from the inlet to the lower border of the cricoid cartilage. It is divided into three parts by two pairs of anteroposterior horizontal folds projecting into the cavity, the upper *vestibular* and lower *vocal folds (false and true vocal cords)*. The part above the vestibular folds is called the *vestibule*. The space between the vocal folds is the *glottis* which is bounded by the vocal folds and extends backwards between the arytenoid cartilages (*rima glottidis*).

The whole lining is covered by mucous membrane; the vocal folds are covered by stratified squamous epithelium; the rest of the lining is covered by ciliated mucous columnar epithelium.

The *vocal fold (true vocal cord)*, one on each side, is attached in front to the angle of the thyroid cartilage and behind to the vocal process of the arytenoid cartilage. It consists of the upper free edge of the cricovocal membrane (vocal ligament) and the vocalis muscle, and is covered by firmly adherent stratified squamous epithelium.

The *ventricle* of the laynx is the fossa between the vestibular and vocal folds; the anterior part of the ventricle is prolonged upwards between the vestibular fold and the lamina of the thyroid cartilage forming a pouch, the *saccule*.

The arteries and nerves of the larynx

The former are the laryngeal branches of the superior and inferior thyroid arteries, the superior supplying the larynx above the vocal folds and the inferior the part below the vocal folds.

The internal laryngeal nerve supplies mucous membrane above vocal folds and the recurrent laryngeal mucous membrane below vocal folds. The recurrent laryngeal nerve supplies all the muscles except the cricothyroid which is supplied by the external laryngeal nerve.

The intrinsic muscles of the larynx

Cricothyroid
Origin: Lateral surface of cricoid cartilage.
Insertion: Lower border of ala of thyroid and front of inferior cornu.
Action: Tensor of vocal folds by pulling thyroid cartilage downwards and forwards or cricoid cartilage upwards and backwards (rotation at cricothyroid joints).

Thyro-arytenoid
Origin: Inside thyroid cartilage from junction of laminae just lateral to attachment of fold. Medial part (*vocalis muscle*) lies in vocal fold.
Insertion: Lateral surface of arytenoid.
Action: Reduces tension in and adducts vocal fold. Vocalis reduces operating length of vocal fold and alters its thickness.

Posterior crico-arytenoid
Origin: Oval area on either side of midline on back of cricoid.
Insertion: Muscular process of arytenoid.
Action: Abductor of vocal folds. This is chiefly by drawing arytenoids downwards and laterally (and therefore apart) along the sloping facets on upper lateral angle of cricoid lamina; very slight rotary effect on arytenoids aids in enlarging glottis. In many animals rotary movement predominates. Also tenses vocal folds.

Lateral crico-arytenoid
Origin: Upper border of lateral part of cricoid arch.
Insertion: Muscular process of arytenoid.
Action: Adduction of vocal folds by rotation forwards of muscular process so that vocal process moves inwards.

Transverse arytenoid
Origin: Back of one arytenoid.
Insertion: Back of other arytenoid.

Action: Approximates arytenoids thus adducting the vocal folds.

Oblique arytenoid
Posterior to transverse arytenoid.
Origin: Back of arytenoid. Fibres pass obliquely to opposite side into aryepiglottic fold to form *aryepiglottic muscle*.
Insertion: Side of epiglottis on opposite side, and to corniculate and cuneiform cartilages.
Action: Sphincter of opening; closes aperture of larynx by approximating arytenoids and aryepiglottic folds to tubercle of epiglottis.

All these muscles are supplied by nerve fibres whose cell bodies are in the *nucleus ambiguus*. The fibres leave mainly in the cranial accessory which joins the vagus. The *superior laryngeal* branch of the vagus divides into the *external* and *internal laryngeal nerves*. The former supplies the cricothyroid, and the *recurrent laryngeal nerve* from the vagus supplies the remaining muscles.

Summary of action of the muscles
(1) Vocal folds tensed by cricothyroids and relaxed by thyro-arytenoids.
(2) Vocal folds separated (abducted) by posterior crico-arytenoids. Vocal folds approximated by lateral crico-arytenoids, transverse arytenoid and thyro-arytenoids.
(3) Inlet of larynx closed by aryepiglottic muscles.

THE TRACHEA AND PRINCIPAL BRONCHI

Trachea
Extends from the lower border of the cricoid cartilage at level of the 6th cervical vertebra to the 5th thoracic vertebra, where it divides into the two principal bronchi slightly to right of midline. It is about 10–12 cm long.
The tube is made of fibro-elastic tissue and is to some extent extensile. It is prevented from collapsing during

inspiration by sixteen to twenty hyaline cartilages which are U-shaped, open posteriorly; the posterior free ends are connected by smooth muscle (*tracheal muscle*).

The trachea is lined by respiratory mucous membrane containing mucous and serous glands and has a ciliated columnar epithelium.

Blood supply is by the inferior thyroid artery, which carries sympathetic fibres from middle cervical ganglion. Parasympathetic supply is via both recurrent laryngeal nerves and right vagus.

Relations: In the neck these are symmetrical. It is clasped laterally by lobes of thyroid gland down to 6th ring, and by the carotid sheaths more distally. Isthmus of thyroid gland adherent over 2nd, 3rd and 4th tracheal rings. Pretracheal fascia, containing inferior thyroid plexus of veins, covers it inferior to isthmus. Posteriorly lies the oesophagus, with recurrent laryngeal nerves in the groove between the two. *In the thorax*, the relations are asymmetrical. On left, arch of aorta and its branches (left common carotid and subclavian) are lateral. On right, vagus and right apex of lung are lateral; more inferiorly azygos vein arches over hilum of lung to enter superior vena cava. Left recurrent laryngeal nerve in tracheo-oesophageal groove. Oesophagus lies in contact with posterior surface in both neck and superior mediastinum.

Right bronchus

About 4 cm long; shorter, more vertical in direction and of larger calibre than the left. It passes to the hilum of the corresponding lung, lying at first above and then behind the right pulmonary artery. Before reaching the pulmonary artery it gives off the upper lobe bronchus to the upper lobe of the right lung, and the vena azygos arches forwards above it.

Left bronchus

About 6 cm long; passes to left under the arch of the aorta to the hilum of the left lung at the level of the upper border of 6th thoracic vertebra. Behind it are the oesophagus, thoracic duct and descending aorta. It lies at first behind the bifurcation of the pulmonary trunk and then below the left pulmonary artery.

The structure of the bronchi is similar to that of the trachea but the cartilage is in the form of short spirals and the muscle is more irregular in its arrangement.

THE LUNGS AND THE PLEURAE

The two lungs occupy the thorax except the mediastinum, which separates them from each other. They are conical in shape and are covered with pleura. Each lung is free except at the median part or root where the blood vessels and bronchi enter and leave.

THE PLEURAE

These are serous sacs enclosing and investing the lungs. Each pleura consists of *visceral* and *parietal layers* between which is the *pleural cavity* containing a small amount of serous fluid. The visceral layer covers the lung, and the parietal layer lines the inner surface of the chest wall, the upper surface of the diaphragm and the side of the pericardium and superior mediastinum. The visceral layer of pleura becomes continuous with the parietal layer in front of and behind the hilum of the lung; below the root a fold, the *pulmonary ligament*, extends downwards along the medial surface of the lung. It may provide space for distension of the pulmonary veins.

The pleura covering the apex of the lung is called the *dome of the pleura*. The *costodiaphragmatic recess* is the space between the inferior visceral pleura and the diaphragmatic parietal pleura.

Surface marking of pleurae
The limits of the parietal pleurae are as follows: Each extends upwards above the medial end of the clavicle for about 3 cm. This is because of the obliquity of the thoracic inlet due to the 1st rib sloping downwards and forwards from behind. Each pleura passes downwards and forwards

to the posterior aspect of the sternoclavicular joint and meets its fellow in the midline at the manubriosternal joint; they pass down together to level of 4th costal cartilages, where the right pleura passes vertically to level of 6th right costal cartilage in midline; then laterally, crossing 10th rib in mid-axillary line; then backwards along the 11th rib to neck of 12th rib. The left pleura at level of 4th costal cartilage arches laterally leaving part of anterior surface of pericardium uncovered. It lies about 2 cm from the left margin of sternum to reach 7th left costal cartilage below which it follows same line as on right, but is placed at a slightly lower level.

THE LUNGS (Fig. 75)

Each lung is cone-shaped with the base inferior.

The *apex* projects upwards behind the medial one-third of clavicle and anterior scalene muscle. Above the first rib, the 1st part of the subclavian artery lies anteriorly separated from it by the pleura.

The *base* is concave, resting upon the diaphragm.

The *costal surface* is convex and corresponds to the chest wall.

The *medial (mediastinal) surface* is concave, corresponding in part to the convex outer surface of the pericardium. It presents about its middle, and towards the posterior part, the *hilum* of the lung where the bronchi and vessels pass in and out to form the *root*. The structures related to this surface are shown in Fig. 75.

The *anterior border* is thin and overlaps the pericardium. On the left lung, from the level of the 4th costal cartilage, it deviates to the left for about 3 cm (*cardiac notch*).

The *posterior border* is rounded and occupies the groove by the side of the vertebrae.

Fissures and lobes

The *left lung* is smaller and narrower than the right, and is divided into an *upper* and *lower lobe* by the *oblique fissure*, which passes upwards and backwards from the hilum on the mediastinal surface on to the costal surface on which it passes downwards and forwards along the line of the 6th

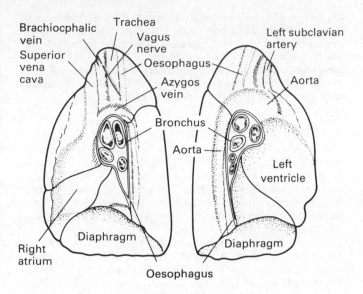

Fig. 75 Mediastinal surface of right and left lungs.

rib to the lower border. It then passes upwards to the hilum. A small tongue of upper lobe between this fissure and the cardiac notch is known as the *lingula*. It corresponds with the middle lobe of the right lung. The *right lung* is divided into three *lobes* (*upper*, *middle* and *lower*) by two fissures. The oblique fissure is similar to the oblique fissure of the left lung. The second (*horizontal fissure*) passes from the middle of the oblique forwards to the anterior border and divides the upper lobe into upper and middle lobes.

The *hilum* is where the bronchus, pulmonary artery and pulmonary veins enter and leave and form the *root*. Here also lie the pulmonary nerve plexuses, bronchial arteries and veins, and the hilar lymph nodes. Disposition of bronchi and pulmonary vessels on cut surface depends on level of section (Fig. 76). On right, bronchus to upper lobe stems from right main bronchus before it enters the lung.

On left, undivided bronchus enters the lung. Pulmonary
artery lies in front of bronchus. Upper and lower pulmonary veins vary in their position but usually lie below the
artery and bronchus. The structures of the root of the right
lung pass behind the ascending aorta. Those of the left pass
in front of the descending aorta where the recurrent
laryngeal nerve hooks round the aortic arch.

The bronchial tree (Fig. 76)

The *principal (main, primary) bronchus* goes to a lung and
gives off *lobar (secondary) bronchi*. Right upper lobe bronchus
branches from main bronchus outside lung. Middle lobe
bronchus branches off within the lung. In left lung, upper
lobe bronchus divides into two and the lower division goes
to *lingular lobe* within the lung (lingular lobe is counterpart
of right middle lobe).

Each lobar bronchus gives off *segmental (tertiary) bronchi*.
Upper lobe has three segments, their bronchi on the right
being *apical*, *posterior* and *anterior*, all coming off together.
On the left, upper lobe bronchus divides into *apicoposterior*
and *anterior*, former quickly dividing into *apical* and *posterior*. *Middle lobe* has two segments. On the right, these are
lateral and *medial*. On the left, lingular lobe has *superior* and
inferior segments. *Lower lobe* has five segments. *Apical bron-*

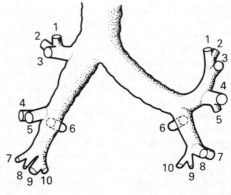

Fig. 76 Segmental
bronchi.

chus to lower lobe is first posterior branch from bronchial tree (inspired liquid, in recumbency, enters this bronchus). Lower lobe bronchus then divides into *medial* (small on left), *anterior*, *lateral* and *posterior* basal segments.

Nerve supply of lungs (p. 301).

THE MEDIASTINUM (Fig. 77)

The mediastinum is, strictly speaking, the septum between the two lungs, but the name is used for the space between the two pleural sacs; it extends from the sternum in front to the thoracic vertebrae behind, and from the thoracic inlet above to the diaphragm below.

For descriptive purposes it is divided into a *superior* and an *inferior mediastinum*, and the latter is again subdivided into *anterior*, *middle* and *posterior*.

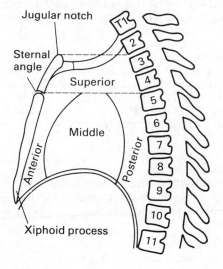

Fig. 77 Subdivisions of mediastinum.

Superior mediastinum
Space above the pericardium extending upwards to the root of the neck.

Boundaries: Superior—thoracic inlet. Inferior—plane passing from lower border of body of 4th thoracic vertebra to the manubriosternal joint. Anterior—manubrium and origins of sternohyoid and sternothyroid muscles. Posterior—first four thoracic vertebrae. Laterally—the parietal (mediastinal) pleura.

Contents: In midline the trachea (lower end) in front of oesophagus; thoracic duct behind left of oesophagus; arch of aorta passing backwards to left with its large branches (brachiocephalic, left common carotid and left subclavian) passing upwards; left brachiocephalic vein immediately above arch of aorta passing to right, joining right brachiocephalic vein to form superior vena cava; right phrenic nerve posterolateral to right brachiocephalic vein, right vagus nerve to right of trachea, left phrenic and vagus nerves to left of arch of aorta; left recurrent laryngeal nerve between oesophagus and trachea; cardiac nerves from neck; lymph nodes and remains of thymus behind sternum.

Anterior mediastinum
Space between the two pleurae in front of the pericardium; it widens inferiorly. It is bounded anteriorly by the sternum and 5th, 6th and 7th costal cartilages, transversus thoracis muscle and left internal thoracic vessels, and posteriorly by the pericardium. It contains lymph nodes and remains of thymus.

Middle mediastinum
Contains pericardium and its contents, phrenic nerves and accompanying vessels, arch of azygos vein, the roots of the lungs and bronchial lymph nodes.

Posterior mediastinum
Lies behind pericardium and diaphragm.

Boundaries: Anterior—pericardium and roots of lungs,

and inferiorly the diaphragm. Posterior—vertebral column, from the lower border of the 4th to the 12th thoracic vertebra. Laterally—pleura. Inferior—diaphragm. Superior—plane passing forwards from lower border of 4th thoracic vertebra.

Contents: Descending thoracic aorta (to left of vertebrae and then in midline), some of its right intercostal branches, oesophagus (in midline and then to left in front of aorta) with vagus nerves, splanchnic nerves passing downwards through diaphragm, the azygos and hemi-azygos veins, thoracic duct to right of oesophagus and posterior mediastinal lymph nodes.

11
The Urogenital System

The urinary and genital organs are developmentally related to each other and in the male both sets of organs have a common terminal channel to the exterior.

THE URINARY ORGANS

THE KIDNEYS (Fig. 78)

The kidneys, which excrete the urine, are situated behind the peritoneum. They extend from the 11th rib to within

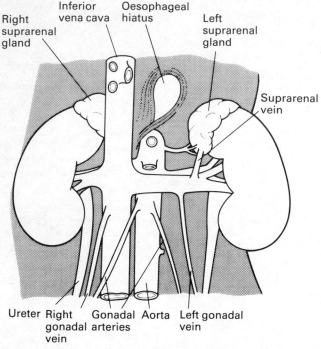

Fig. 78 Kidneys on posterior abdominal wall.

3 cm of the iliac crest, the right being lower than the left. The average length of each kidney is 10 cm, breadth 6 cm and thickness 3 cm. Each weighs about 150 g.

Relations
The anterior surface faces laterally and the upper pole is nearer the midline than the lower; each kidney is partly

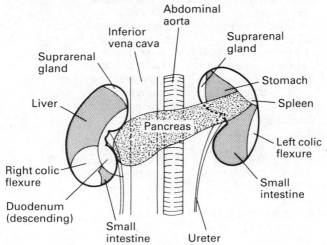

Fig. 79 Anterior relations of kidneys.

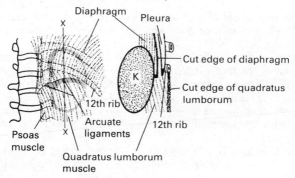

Fig. 80 Posterior relations of kidneys.

covered on its anterior surface by peritoneum and the relations of the two kidneys differ.

Right kidney:

Anterior (Fig. 79)	*Posterior* (Fig. 80)
Right lobe of liver	Diaphragm, separating it
Descending (2nd) part of	from pleura
duodenum	12th rib
Right colic flexure	Transversus abdominis
Coils of jejunum	Quadratus lumborum
(No peritoneum between	Psoas major
kidney and duodenum	Lumbar fascia
and colon)	Subcostal vessels and nerve
	Iliohypogastric nerve
	Ilio-inguinal nerve

Left kidney:

Anterior	*Posterior*
Fundus of stomach	As on right but left
Posteromedial surface of	kidney is higher and
spleen	reaches 11th rib
Left colic flexure	
Jejunum	
(No peritoneum between	
kidney and pancreas	
and colon)	

At upper pole of each kidney there is a suprarenal gland.

The lateral border is convex and is directed laterally and backwards.

The medial border, concave and directed medially and forwards, contains hilum where the artery enters and the vein and ureter leave; from before backwards there are the renal vein, renal artery, ureter. Renal artery usually divides and enters hilum as three vessels, two in front and one high up behind ureter. *Aberrant artery* from aorta to lower pole very common.

Each kidney has its own *fibrous capsule* and is embedded in *perirenal fat (adipose capsule)* which is contained within the *renal fascia* (part of the fascia transversalis).

THE URETERS (Fig. 81)

Each kidney is connected with the bladder by a ureter, 25 cm long, which conveys urine to the latter viscus. The upper end of each ureter is expanded, the *pelvis* now called the *renal pelvis* (capacity about 5 ml). The pelvis is divided

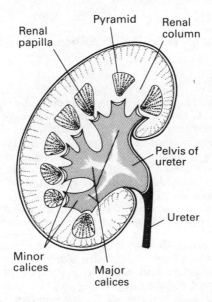

Fig. 81 Vertical section through medial and lateral borders of kidney.

into two or three *greater calices*; these are subdivided into about *twelve lesser calices*. Into these calices small *papillae* project, which are the apices of the *renal pyramids*. These latter form the *medulla* of the kidney and are embedded in the *cortex* which is peripheral to the medulla and extends inwards between the pyramids as the *renal columns*

Relations in abdomen

Anterior	*Posterior*
Peritoneum	Psoas
Colic vessels	Genitofemoral nerve
Gonadal vessels	Common or external iliac
Ileum and mesentery	artery
(right side)	
Pelvic colon and mesocolon	
(left side)	

The right ureter lies close to the lateral side of the inferior vena cava.

Relations in pelvis

In the male: Each ureter enters the pelvis by crossing the bifurcation of the common iliac artery at level of sacro-iliac joint. It then runs down to the ischial spine, along a line immediately in front of the internal iliac artery. It crosses the obturator nerve and the anterior branches of the internal iliac artery and then turns medially by passing below the ductus deferens before entering the bladder.

Note: Surface marking (for radiographic identification): 2nd lumbar vertebra, a line along the tips of the lumbar transverse processes, sacro-iliac joint at brim of pelvis, spine of the ischium, pubic tubercle.

In the female: Course is as above: the ureter passes forwards about 1 cm lateral to the side of the cervix below the uterine artery and lateral to upper part of vagina to posterior part of base of bladder.

Narrowest parts of ureter are (1) pelvi-ureteric junction (2) brim of pelvis (3) at entrance to bladder. Calculus most frequently impacted at one of these places.

THE BLADDER (Fig. 82)

Position

In infancy it lies mainly in abdomen. In adult it lies in pelvis behind pubis; in the male it is in front of rectum and in the female it is in front of uterus and vagina.

When empty, the bladder is described as having four

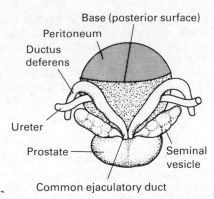

Base (posterior surface)
Peritoneum
Ductus
deferens
Ureter
Prostate
Seminal
vesicle
Common ejaculatory duct

Fig. 82 Base of
bladder.

triangular surfaces meeting anteriorly at the *apex*. Postero-inferiorly is the *base*. There are two antero-inferior surfaces and a superior surface. As bladder enlarges it becomes more globular and extends upwards into abdomen, anterior to the peritoneum, so that it is directly in contact with anterior abdominal wall.

The apex is connected to the umbilicus by the *median umbilical ligament (urachus)* and by the *medial umbilical ligaments* (obliterated umbilical artery).

Relations
The *body* (between apex and base) is behind the symphysis and body of the pubis, separated by retropubic space containing fat, plexus of veins and puboprostatic ligaments (pubovesical in female). Upper surface is covered by peritoneum and is in contact with pelvic colon in male and uterus in female and with small intestine in both sexes. Passing obliquely on the side of the bladder when distended is the *medial umbilical ligament*, which marks the lateral limit of its peritoneal covering; below this the side of the bladder is separated by loose connective tissue from the obturator internus and levator ani. The ductus deferens passes obliquely along the lower part of the lateral surface medial to the ureter and medial umbilical ligament.

The *base* is directed backwards and downwards. In the male, it is separated from the rectum by the rectovesical pouch of peritoneum below which there is the rectovesical fascia. On each side the ductus deferens is medial to the seminal vesicle; these unite to form the (common) ejaculatory duct. Duct enters prostate which is inferior to base at *neck* of bladder (inferior angle of base where bladder is continuous with the prostatic urethra). In female, bladder base is separated from uterus and anterior wall of vagina by uterovesical pouch.

Ligaments

True ligaments: Presumed to support the bladder and are thickenings of pelvic fascia. There are two *lateral* (one on each side) from bladder to pelvic wall. In the male there are two anteriorly on each side (lateral and medial) called *puboprostatic* (in female *pubovesical*), from region of symphysis to prostate (or bladder). *Posterior ligament* is fascia round plexus of veins passing backwards to internal iliac vein.

False ligaments: Folds of peritoneum; two posterior (*sacrogenital folds*); two *lateral* (from bladder to lateral pelvic wall); anteriorly one *median umbilical fold* covering the urachus and two *medial umbilical folds* covering the obliterated umbilical arteries (p. 360).

Interior of the bladder

On the inner surface of the base of the bladder just behind the urethral orifice there is a triangular smooth area or *trigone* with the apex anterior and inferior. It is bounded by the ureteric orifices above and laterally, and by the opening of the urethra inferiorly. At its apex behind the urethra there is an elevation formed by the prostate, called the *uvula of bladder* which is about 3 cm from the ureters. Over the trigone the mucous membrane is smooth and paler than that of the rest of the bladder; elsewhere the lining is ridged. The ridges decrease with filling of the bladder. The *interureteric crest* is between the ureteric orifices which are about 3 cm apart.

The male *urethra* is described on p. 388 and the female on p. 394.

THE MALE GENITAL ORGANS

THE PROSTATE

The prostate surrounds the neck of the bladder and the first part of the urethra, which is placed nearer the anterior than the posterior surface of the gland. It resembles a horse-chestnut in shape, with the apex directed downwards. It measures about 3 cm transversely, 3 cm high and 2.5 cm from before backwards. It is held in position by the puboprostatic ligaments.

The prostate is perforated by the (common) ejaculatory ducts (p. 392) which open into the floor of the prostatic urethra.

Relations

Anterior: Symphysis pubis, puboprostatic ligaments, retropubic fat with plexus of veins and dorsal vein of penis.

Posterior: Rectum, rectovesical fascia (of Denonvilliers).

Superiorly: Base of bladder with ductus deferentes and seminal vesicles.

Inferiorly: Pelvic fascia between diverging anterior edges of levatores ani.

The prostate is surrounded by a sheath of pelvic fascia within which is the prostatic plexus of veins outside its own capsule. It is supported by the levator ani.

In addition to the lobules of the prostate, the ducts of which open into the prostatic urethra (p. 389), lobes of the prostate are described. The *right* and *left lateral lobes* are demarcated by a vertical groove on its posterior surface and the *middle (median) lobe* lies between the prostatic urethra and ejaculatory ducts.

THE BULBO-URETHRAL GLANDS

These are two small, round bodies about the size of a pea, lying behind the membranous urethra, between the perineal membrane and the parietal pelvic fascia (superior fascia of the urogenital diaphragm) in the deep perineal pouch. Their ducts, about 2.5 cm long, pass forwards and pierce the perineal membrane to open into the spongy part of the urethra in the superficial perineal pouch.

THE PENIS *(also pp. 128–9)*

The penis is divided into a *root*, *body* and *glans*.

The *root* is attached to the perineal membrane (inferior fascia of the urogenital diaphragm), to the ischiopubic rami by the crura, and to the symphysis pubis by the suspensory ligament.

The *glans* forms the extremity; at its summit is the opening of the urethra, the *external meatus*. The *frenulum*, a fold of mucous membrane, passes from the lower margin of the meatus to the *prepuce*. The base of the glans projects to form the *corona*, behind which there is a constriction, the *neck of the glans*; sebaceous glands are found on both. The skin of the penis, attached to the neck of the glans, is doubled upon itself to form the *prepuce* or *foreskin*.

The *body* is the part between the root and the glans and consists of the *corpora cavernosa* and *corpus spongiosum*; its upper surface is the *dorsum*.

The *corpora cavernosa*, placed dorsally, are two cylindrical columns connected together for their distal three-fourths, with the *septum of penis* intervening. As they approach the root they separate to form the two *crura*, which are attached to the medial margins of the ischiopubic rami. Each crus is surrounded by a muscle, the *ischiocavernosus*. Anteriorly the corpora cavernosa fit into the base of the glans. There is a groove on the superior surface for the dorsal vein of the penis and another groove on the inferior surface for the corpus spongiosum; the corpora cavernosa are attached to the pubic symphysis by the *suspensory ligament*.

The *corpus spongiosum* commences at the perineal membrane as an enlargement, the *bulb*, runs forward in the groove on the inferior surface of the corpora cavernosa, and expands over their extremities to form the *glans*. The bulb is embraced by the *bulbospongiosus muscle*. The urethra pierces the bulb on its upper surface, and then runs forward in the middle of the corpus spongiosum to end at the external meatus.

THE MALE URETHRA

The urethra in the male extends from the neck of the

bladder to the end of the penis, and has a length of from 15 to 20 cm. It is divided into three parts according to the structure through which it passes.

(1) The *prostatic part*, widest and most dilatable, passes through the prostate gland from base to apex; this part is about 3 cm long and spindle-shaped; in cross-section it is horseshoe-shaped, with the convexity forwards. On the posterior wall is a longitudinal ridge, the *urethral crest*, and on each side of this promontory is a depression, the *prostatic sinus*, into which the prostatic ducts open. Towards the anterior part of the urethral crest is a depression, the *prostatic utricle*, upon the elevated edges of which the ejaculatory ducts open.

(2) The *membranous part*, narrowest and least dilatable, extends from the apex of the prostate to the bulb, and is 2 cm long; it is contained between the perineal membrane and the pelvic fascia (the two layers of the urogenital diaphragm) and is surrounded by the sphincter urethrae (p. 129).

(3) The *spongy part* is in the corpus spongiosum and glans and is about 10–15 cm long; the part in the bulb is somewhat dilated and the ducts of the bulbo-urethral glands open on the floor; the canal enlarges again before ending at the external meatus (*fossa navicularis*). The lumen of this part of the urethra is transverse, except at the meatus (its narrowest part) where it is vertical, hence the spiral stream of urine.

The floor of the urethra contains depressions (*lacunae*) on to which urethral glands open in a forward direction and an instrument passed into the urethra is liable to catch the edge of the lacuna; one large one in the fossa navicularis is called the *lacuna magna*.

THE SCROTUM

The scrotum is the bag containing the testes each of which is suspended by a spermatic cord. The skin has a median raphe and a fibrous septum divides the scrotum into two cavities; the left half is longer than the right because the left testis is lower and the left spermatic cord longer than on the right.

Coverings of the testis in the scrotum
(1) skin ⎫ Common to
(2) superficial fascia and dartos muscle ⎬ both testes
(3) external spermatic fascia ⎭
(4) cremaster muscle and fascia
(5) internal spermatic fascia, continuous with the transversalis fascia
(6) tunica vaginalis testis; forms a closed sac and invests the body and epididymis of the testis except posteriorly where the duct and vessels are attached; laterally it passes a short distance between epididymis and body and forms the *sinus of the epididymis.*

THE SPERMATIC CORD

The spermatic cord consists of the *ductus deferens* with *artery to the ductus, testicular artery* and *pampiniform plexus of veins* (forming testicular vein above), *sympathetic nerves*, the *artery to cremaster*, the *genital branch of the genitofemoral nerve, lymphatics* and some *areolar tissue*; it extends from the deep inguinal ring to the testis, passing in its course through the inguinal canal, from which it emerges at the superficial inguinal ring, and then in front of the pubis to the scrotum. The ductus deferens is at the back of the cord, and may be recognized by its hard and cord-like feeling.

Boundaries of the inguinal canal (p. 119).

THE TESTES (Fig. 83)

Each testis is suspended in the scrotum by the spermatic cord which is attached to its posterior border. A testis consists of two parts, the *body* which is anterior, and the *epididymis* which is posterior. The ductus deferens is attached to the lower end of the latter.

Coverings of the testis
(1) *Tunica vaginalis* (serous) derived from the peritoneum; tunica vaginalis is attached to the underlying tunica albuginea.

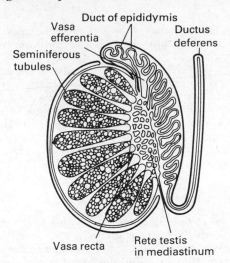

Fig. 83   Vertical section through testis.

(2) *Tunica albuginea* (fibrous) covers the body of the testis; thickened posteriorly to form a vertical septum, the *mediastinum testis* which gives off septa separating the lobules of the testicle.

Epididymis
A long narrow structure consisting of three parts, a superior *head* projecting forwards at its upper end, a lower tail with which the ductus deferens is continuous and between these two, the *body*. The epididymis consists of a very convoluted duct, 6–7 m long, bound together by areolar tissue; it receives the *efferent ductules* from the upper part of the testis.

Body of testis
Consists of numerous pyramidal lobules which are formed by septa from the mediastinum testis; the base of the lobule is directed towards the circumference of the testis and the apex towards the mediastinum. Each lobule contains one to three convoluted *seminiferous tubules* held together by areolar tissue. A tubule commences near the base, in a blind

extremity and becomes straighter near the apex; as they enter the mediastinum, they coalesce and form twenty or thirty *straight tubules*.

The straight tubules pierce the mediastinum and interlace to form the *rete testis*.

Twelve to twenty tubules, *efferent tubules*, pass from the upper part of the rete and perforate the tunica albuginea. They are at first straight and then convoluted. They form the *lobules of the epididymis* in its head.

The lobules of the epididymis open into the excretory duct (when unravelled) at intervals of about 7 cm.

THE DUCTUS DEFERENS

Commences at the tail of the epididymis as the continuation of the duct of the epididymis and ascends in the spermatic cord through the inguinal canal and deep inguinal ring. Here it hooks round the lateral side of the inferior epigastric artery. It is external to the peritoneum and passes medially to reach the medial side of the external iliac artery. It then lies on the lateral pelvic wall and crosses medial to the obliterated umbilical artery, obturator nerve and vessels, vesical vessels and ureter. At the base of the bladder it is medial to the seminal vesicles inferior to the peritoneum. Behind the bladder it is dilated to form the ampulla and is separated from the rectum by the rectovesical fascia. It narrows again at the base of the prostate and unites with the duct of the seminal vesicle of the same side to form the *(common) ejaculatory duct,* which passes forwards through the prostate and opens into the prostatic urethra on the edge of the orifice of the prostatic utricle.

THE SEMINAL VESICLES

These are two sacculated pouches lying between the base of the bladder in front and the rectum behind. They are pyramidal in shape, the upper end being the wider; inferiorly they converge and join the corresponding ductus deferens which lies on its medial side to form the ejaculatory duct.

THE FEMALE GENITAL ORGANS

THE EXTERNAL ORGANS

These include the mons pubis, labia majora, labia minora, clitoris, external orifice of the urethra and orifice of the vagina. The term *vulva* includes all of these.

Mons pubis

Eminence in front of the pubis due to underlying fat; covered with hair.

Labia majora

Two prominent folds extending from the mons to the perineum, the region between the vagina and anus. Externally they are covered with hair and skin, internally with mucous membrane; the labia are joined together anteriorly and posteriorly, forming commissures. A small transverse fold of mucous membrane is found in the posterior commissure called the *frenulum labiorum*; the space between this and the posterior commissure is the *vestibular fossa*.

Labia minora

Two folds of mucous membrane extending for 3 cm downwards and laterally from the clitoris, finally losing themselves posteriorly in the labia majora. Anteriorly they divide and surround the clitoris, the anterior folds forming the *prepuce of the clitoris*, the posterior, attached to the glans, forming the *frenulum of the clitoris*.

Clitoris

Corresponds to the penis; lies just behind the anterior commissure. It consists of two *corpora cavernosa* attached to the pubic rami by two *crura*; the free extremity or *glans* also consists of vascular erectile tissue. Between the labia minora, and bounded anteriorly by the clitoris, is the *vestibule*; the *external urethral orifice* opens into the vestibule about 3 cm behind the clitoris.

Bulbs of the vestibule

Masses of erectile tissue lying deep to the labia minora on

each side of the vagina; attached to the inferior surface of the perineal membrane.

Behind the orifice is the *opening of the vagina*, narrowed in the virgin by the *hymen*, a duplicature of mucous membrane, generally semilunar in shape. After its rupture small elevations, *hymeneal caruncles*, remain.

Greater vestibular glands
Analogous to bulbo-urethral glands in the male; situated on each side near the entrance of the vagina, with ducts opening on the labia minora, external to the hymen.

Urethra (in the female)
About 3 cm long; is adherent to the anterior wall of the vagina. The external orifice of the urethra opens between the labia minora, about 3 cm behind the clitoris.

Vagina
Dilatable canal with the anterior and posterior walls approximated to each other so that the lumen is a transverse slit; extends from the vulva to the uterus; the anterior wall is about 8 cm and the posterior wall about 11 cm long. The upper end widens to receive the cervix of the uterus which it meets at an angle anteriorly of about 90°.

It is anterior to the rectum and anal canal. Its opening is separated from the anal canal and anus by the *central tendon of the perineum (perineal body)*. The lowest part of the recto-uterine pouch is related to its upper one-third. The base of the bladder and the urethra are anterior to the vagina. The lower part of the broad ligament and its connective tissue containing the ureter are lateral.

There is a median ridge or raphe on the mucous surface of both the anterior and posterior walls. The walls likewise present many transverse ridges or rugae. The lower end of the vagina is embraced by the sphincter vaginae (p. 126), part of the levatores ani.

THE INTERNAL ORGANS
The uterus (Fig. 84)
The uterus is pear-shaped, flattened from before backwards and placed in the pelvis between the bladder and the

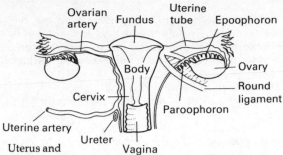

Fig. 84 Uterus and broad ligament.

rectum. The nulliparous adult uterus is about 7 cm long with its long axis directed forwards and upwards in the line of the axis of the pelvis. Peritoneum covers the back of the uterus and extends down on to the upper third of the vagina; anteriorly only the upper two-thirds of the uterus is covered. The peritoneum is reflected from the sides of the uterus to the lateral wall of the pelvis, forming the *broad ligaments*. The uterus is divided into: (1) the *fundus*, the broad upper end, convex in both directions and covered by peritoneum; (2) the *body* extending from the fundus to the cervix (neck); narrows as it approaches the latter; uterine tube is attached at the junction of fundus and body (*cornu*); the round ligament of the uterus (anterior), and the ovarian ligament (posterior) are attached at the cornu; (3) the *cervix* of the uterus projects into the vagina—*vaginal cervix*. Part of the cervix is above the vagina—*supravaginal cervix*. The opening of the cervix into the vagina (*os uteri*) is bounded by two thick lips, anterior and posterior, of which the anterior is the thicker and the posterior the longer. The recess between the vaginal cervix and vault of the vagina is called the *fornix* which is divided into an anterior and posterior and two lateral fornices. The posterior is the deepest and is related to the peritoneum of the recto-uterine pouch.

The *cavity* of the uterus, flattened from before backwards, is triangular in shape; the superior angles lead to the uterine tubes.

The narrow part of the uterus between the body and the cervix is called the *isthmus* of the uterus.

Usually the body is bent forwards on the cervix (*ante-flexion*) and the whole uterus bent forwards on the vagina (*anteversion*). The uterus lies on the superior surface of the bladder but becomes more vertical as the bladder fills.

The ligaments of the uterus

These are usually divided into two groups: (a) the broad ligament and round and ovarian ligaments, (b) fascial ligaments of which there are three pairs (lateral, anterior and posterior) and are regarded as the important structures for maintaining the position of the uterus; they are related to its cervix.

The *broad ligament* consists of a double layer of peritoneum passing from the lateral margin of the uterus to the side of the pelvis. Between the two layers of this ligament on each side are the uterine tube in the medial four-fifths of the upper border, the round ligament deep to its anterior layer, fetal relics, the *epoophoron* (lateral) and *paroophoron* (medial), the ovary and its ligament related to its posterior layer, uterine and ovarian vessels, nerves and lymphatics. Inferior to the broad ligaments there is thickened fascia related to the cervix. In its upper part the uterine artery passes to the uterus and the ureter passes forwards inferior to the artery to the bladder.

The *round ligament* is a cord between the layers of the broad ligament, extending from the cornu of the uterus to the deep inguinal ring; thence it passes into the inguinal canal into the labium majus. In the fetus it is accompanied for some part of the way in the inguinal canal by peritoneum, the *processus vaginalis*; this process may persist.

The *ovarian ligament* passes from the lower (medial) pole of the ovary to the cornu of the uterus and raises a ridge on the posterior layer of the broad ligament. It is continuous below the uterine tube with the round ligament. Both are remains of the *gubernaculum ovarii*.

The *fascial ligaments* are (1) the *lateral cervical (cardinal) ligaments* from the side of the cervix to the lateral wall of the pelvis, (2) the *pubocervical* from the body of the pubis to the cervix, (3) the *uterosacral* from the sacrum round the sides of the rectum to the cervix. These are frequently torn during labour and the result is a descent of the uterus and the vault of the vagina (*prolapse of uterus*).

The uterine (fallopian) tubes

The uterine tubes, one on each side, lie in the upper margins of the broad ligaments, and are about 10 cm long. The lateral end of each tube has a small orifice called the *abdominal opening* which lies at the bottom of a trumpet-shaped expansion, the *infundibulum*. The margin of the infundibulum has a number of irregular processes, the *fimbriae*.

Where the tube joins the uterus it is narrow and termed the *isthmus*; between this and the abdominal opening there is a dilatation, the *ampulla*, extending rather more than half the length of the tube.

The fimbriated extremity is closely applied to the tubal end and free border of the ovary, and is attached by the *ovarian fimbria* to the tubal end of the ovary.

The ovaries

The two ovaries correspond to the testes in the male; they are flattened and ovoid and are vertically placed on the posterior surface of the broad ligament. The anterior border of the ovary is attached to the broad ligament by the *mesovarium* containing the ovarian vessels, nerves and lymphatics; the posterior border is free. The upper pole is embraced by the uterine tube; the lower pole is attached to the upper angle of the uterus by the ovarian ligament. The medial surface is free; the lateral may lie in the *ovarian fossa* between the internal and external iliac vessels just in front of the ureter on the obturator nerve.

THE BREAST

The breast (*mammary gland*) is rudimentary in the male and female child. It enlarges in the female at puberty. During lactation it enlarges further. The developed gland consists of 15 to 20 lobules separated by fibrous septa that radiate from the nipple to the deep fascia on pectoralis major (its epimysium). Each lobule opens by a separate *duct* on the *nipple*, and is dilated to form the *lactiferous sinus* just below the surface of the nipple. The nipple is surrounded by a pink *areola* which usually becomes permanently pigmented early in the first pregnancy (degree of pigmentation depends on colouring of individual). Contour of breast is due to subcutaneous fat which is absent under areola and nipple. Gland has no capsule. It lies in subcutaneous tissue over pectoralis major and serratus anterior; often has an *axillary tail* passing upwards over latter muscle and may be deep to deep fascia. Breast overlies 2nd to 6th ribs, from side of sternum to anterior axillary line.

Blood supply: Lateral thoracic, perforating branches of intercostals and internal thoracic, especially through 2nd and 3rd spaces.

Lymphatic drainage: Very important (p. 207).

12
The Ductless Glands

THE THYROID GLAND

This gland is situated on the lower part of the neck and consists of two *lateral lobes* on either side of the midline united in its lower part by an *isthmus*, which lies over the 2nd, 3rd and 4th rings of trachea. It is covered by skin, superficial and cervical fasciae and the anterior jugular veins and is enveloped in pretracheal fascia which strips freely except at the poles where vessels perforate it, and between the isthmus and trachea. Hence the gland moves upwards and downwards with larynx in swallowing.

Each lobe is pear-shaped with the smaller end upwards and is about 3 cm long, 2 cm wide and 1 cm thick.

Relations of lobes
Anterolateral surface: Cervical fascia, sternocleidomastoid, sternothyroid, sternohyoid and omohyoid muscles.
Posteromedial surface: Upper six rings of trachea, cricoid cartilage, thyroid cartilage below oblique line, cricothyroid and inferior constrictor muscles, inferior thyroid artery with recurrent laryngeal nerve, external laryngeal nerve, oesophagus (on left side).
Posterolateral surface: Sheath of carotid vessels.

Vessels
Arteries, superior (from external carotid) and inferior (from thyrocervical trunk) thyroid, thyroidea ima (from aorta); veins, superior and middle (to internal jugular) and inferior thyroid to left brachiocephalic.
Note: The thyroid gland has its own capsule within which there is a plexus of veins and is enclosed in a capsule derived from the pretracheal fascia.

THE PARATHYROID GLANDS

These are two small bodies on each side situated at the back of the lateral lobes of the thyroid gland within its capsule. The *upper* is about the level of the cricoid cartilage, the *lower* near the lower pole of the thyroid. They are supplied by the posterior branches of the inferior thyroid arteries.

Upper is developed from 4th pharyngeal pouch (parathyroid IV), lower from 3rd pouch (parathyroid III) which is pulled caudally during descent of thymus derived from 3rd pouch.

THE SUPRARENAL GLANDS

These are asymmetrical. Right is pyramidal and embraces upper pole of right kidney, left crescentic and embraces medial border of left kidney above the hilum. The right usually has no peritoneum in front, but the left is covered in front by peritoneum, except where the pancreas crosses it.

Relations: Liver and inferior vena cava are anterior to right gland and stomach, lesser sac of peritoneum and pancreas are anterior to left gland. The diaphragm is posterior to both glands.

Arteries: Large; three to each gland—superior suprarenal from inferior phrenic, middle suprarenal from aorta, inferior suprarenal from renal.

Veins, one on each side: Right into inferior vena cava; left into left renal.

Nerves: Very numerous; from splanchnic nerves through coeliac and renal plexuses. Almost entirely preganglionic sympathetic.

PITUITARY GLAND (HYPOPHYSIS CEREBRI, p. 233).

Occupies sella turcica of sphenoid. Fossa roofed by dura mater (*diaphragma sellae*) attached to clinoid processes and pierced by stalk of pituitary (*infundibulum*) which depends from tuber cinereum on floor of 3rd ventricle (part of hypothalamus). Has optic chiasma above, cavernous sinus on each side with internal carotid artery, and sphenoidal sinus below. Anterior lobe larger and more vascular than posterior.

Index

401